CARTESIAN TENSORS

CARTESIAN TENSORS

BY

HAROLD JEFFREYS
M.A., D.Sc., F.R.S.

CAMBRIDGE
AT THE UNIVERSITY PRESS
1957

PUBLISHED BY
THE SYNDICS OF THE CAMBRIDGE UNIVERSITY PRESS

Bentley House, 200 Euston Road, London, N.W. 1
American Branch: 32 East 57th Street, New York 22, N.Y.

First Published 1931
Reprinted 1952
Reprinted 1957

First printed in Great Britain at the University Press, Cambridge
Reprinted by offset-litho by Jarrold & Sons Ltd., Norwich

PREFACE

It is widely felt that when the equations of mathematical physics are written out in full Cartesian form the structural simplicity of the formulae is often hidden by the mechanical labour of writing out every term explicitly. Attempts have been made to reduce this labour by one form or another of vector algebra; but it has always seemed to me that this method both introduces new difficulties and is insufficiently general. Thus the product of two vectors, in vector language, means one of two things, either the scalar or the vector product, and it is not physically obvious why just these functions of the vectors should arise and no others.

The use of tensor notation, with the summation convention, carries out as great a simplification of the writing as does vector notation. The notation has actually attracted attention owing to its applications in the theory of relativity, but for ordinary purposes two great abbreviations may be made. We use rectangular Cartesian axes; the result is that the distinction between covariant and contravariant vectors disappears, and with it the terms arising from curvature of the surfaces of reference. The formidable character of most of the formulae of the theory of relativity is absent from the formulae of tensors referred to Cartesian axes. The tensor method is a necessity for relativity; for applications in dynamics, electricity, elasticity, and hydrodynamics it is a great convenience.

It is found that the scalar and vector products are *not* the only functions of two vectors that arise, though the theory provides reasons why they are important in many applications. There is also a symmetrical product, which ordinary vector notation is completely unable to express. In tensor notation it arises naturally as a symmetrical tensor of the second order. The system of moments and products of inertia of a rigid body constitutes such a tensor; so do the stress components and the strain components in an elastic solid.

The present method, like vector notation, is of use principally in proving general theorems. In concrete applications there is usually some asymmetry about the coordinates that makes it necessary to abandon the tensor form at some stage in the work. It has been said that vector equations are like a pocket map, and it has been replied that a pocket map has to be taken out of the pocket and unfolded before it is of any use. The same applies to the tensor method, and for the same reason; but it has the great advantage that it is not a new notation, but a concise way of writing the ordinary notation, so that the unfolding can be carried out more conveniently when occasion arises.

What is usually called Statics is treated in Chapter v, after Dynamics. I consider this to be the proper order, because Statics is a special case of Dynamics, and many of its formulae have physical significance for reasons explained in Dynamics. The customary reversal of the order is due, I believe, to the fact that an introduction to mechanics has to be given at schools before the students have received any training in calculus; but this need not influence students working for a university examination.

It should perhaps be stated that the object of this work is to illustrate the use of tensor methods; it does not claim to give a complete theory of all the subjects touched, reference for which must be made to the standard text-books.

I must express my gratitude to Mr M. H. A. Newman, Miss L. M. Swain, Dr S. Goldstein, and Dr Bertha Swirles for assistance at various stages in the work, and to the staff of the University Press for their care in the printing.

HAROLD JEFFREYS

September 1931

NOTE

Since this book was written most of the material in it has been incorporated in Chapters 2 and 3 of *Methods of Mathematical Physics*, by my wife, Bertha Swirles Jeffreys, and myself. The present reprint has been made partly because of a continuing demand for a treatment of Cartesian tensors by themselves; partly because some results, notably on the thermodynamics of an elastic solid and the circulation in viscous flow, are not given in textbooks of the special subjects.

HAROLD JEFFREYS

January, 1952

CONTENTS

CHAPTER I

CARTESIAN TENSORS

If we have two sets of rectangular axes (Ox, Oy, Oz), (Ox', Oy', Oz') at the same origin, the coordinates of a point P with respect to the second set are given in terms of the coordinates with respect to the first set by the equations

$$\left.\begin{aligned} x' &= l_1 x + m_1 y + n_1 z \\ y' &= l_2 x + m_2 y + n_2 z \\ z' &= l_3 x + m_3 y + n_3 z \end{aligned}\right\} \qquad (1).$$

The quantities $(l_1, m_1, n_1, \ldots, n_3)$ are the cosines of the angles between the various axes; thus l_1 is the cosine of the angle between the axes Ox' and Ox, n_2 is the cosine of the angle between Oy' and Oz, and so on. It follows that the coordinates (x, y, z) can be expressed in terms of (x', y', z') by the relations

$$\left.\begin{aligned} x &= l_1 x' + l_2 y' + l_3 z' \\ y &= m_1 x' + m_2 y' + m_3 z' \\ z &= n_1 x' + n_2 y' + n_3 z' \end{aligned}\right\} \qquad (2).$$

We can shorten the writing of (1) and (2) considerably by a change of notation. Instead of (x, y, z) let us write (x_1, x_2, x_3), and instead of (x', y', z') write (x_1', x_2', x_3'). We can now say that the coordinates with respect to the first set of axes are x_i, where i may be 1, 2, or 3; and those with respect to the second set are x_j', where j may be 1, 2, or 3. Then in (1) each coordinate x_j' is expressed as the sum of three terms depending on the three x_i. Each x_i is associated with the cosine of the angle between the direction of that

x_i increasing and that of x_j' increasing. Let us denote this cosine by a_{ij}. Then we have, for all values of j,

$$x_j' = a_{1j}x_1 + a_{2j}x_2 + a_{3j}x_3$$
$$= \sum_{i=1,2,3} a_{ij}x_i \qquad (3).$$

Conversely (2) can be written

$$x_i = \sum_{j=1,2,3} a_{ij}x_j' \qquad (4),$$

the a_{ij} having the *same* value as in (3), for the same values of i and j, because it is in both cases the cosine of the angle between the directions of x_i and x_j' increasing.

In mathematical physics we often have to deal with sets of three quantities in relation to a set of axes, of the general form u_i (that is, u_1, u_2, u_3), and such that in relation to a different set of axes the corresponding quantities are (u_1', u_2', u_3'), which satisfy the relations

$$u_j' = \sum_{i=1,2,3} a_{ij}u_i \qquad (5)$$

and

$$u_i = \sum_{j=1,2,3} a_{ij}u_j' \qquad (6).$$

Such sets of three quantities are called *tensors of the first order*, or *vectors*. The individual u_1, u_2, u_3 may be called the *components* of the tensors.

Clearly if we multiply all of the u_i and u_j' by the same quantity m we get

$$mu_j' = \sum_{i=1,2,3} a_{ij}(mu_i) \qquad (7),$$

so that mu_i is another tensor of the first order.

Again, if we have two tensors of the first order, u_i and v_i, we shall have

$$u_j' + v_j' = \sum_{i=1,2,3} a_{ij}(u_i + v_i) \qquad (8),$$

so that $u_i + v_i$ is a tensor of the first order.

We notice that each of the equations (3) to (8) is really a set of three equations; where the suffix i or j appears on the left it is to be given in turn all the values 1, 2, 3, and the resulting equation is asserted in each case. In each such equation the right side is the sum of three terms, obtained by giving j or i the values 1, 2, 3 in turn and adding. Wherever such a summation occurs a suffix is repeated in the expression for the general term; where there is a summation for all values of j the general term, such as $a_{ij}u_j'$, contains j *twice*. We make it a regular convention that, unless the contrary is stated, whenever a suffix is repeated it is to be given all possible values and that the terms are to be added for all. Thus we write (5) as simply

$$u_j' = a_{ij}u_i \qquad (9),$$

the summation sign being automatically understood by our convention. Then (9) really means three equations, with three terms on the right of each, but we can by means of our conventions express all of the twelve terms compactly by the single equation (9).

There are single quantities, such as mass and distance, that are the same for all sets of axes. These are called *tensors of zero order*, or *scalars*.

Consider now two tensors of the first order, u_i and v_k. (When we write "a tensor u_i" we mean of course a tensor of the first order whose components are u_1, u_2, u_3. This is another piece of shorthand.) Suppose each component of the one multiplied by each component of the other; then we obtain a set of nine quantities expressed by $u_i v_k$, where each of i and k is independently given all the values 1, 2, 3. The components of u_i, v_k with respect to the other set of axes are u_j', v_l' say; and

$$u_j' v_l' = (a_{ij}u_i)\,(a_{kl}v_k)$$

$$= a_{ij}a_{kl}u_i v_k \qquad (10).$$

The suffixes i and k are repeated on the right. Thus (10) represents nine equations, each with nine terms on the right. Each term on the right is the product of two factors, one of the form $a_{ij}a_{kl}$, depending only on the orientation of the axes, and the other of the form $u_i v_k$, representing the products of the components referred to the original axes. In this way the various $u_j' v_l'$ can be obtained in terms of the original $u_i v_k$. But products of two vectors are far from being the only quantities satisfying this rule. In general a set of nine quantities w_{ik} referred to a set of axes, and transformed to another set by the rule

$$w_{jl}' = a_{ij}a_{kl}w_{ik} \tag{11},$$

is called a *tensor of the second order*.

We may go on similarly to construct and define tensors of the third, fourth, and higher orders. Thus a set of quantities that transforms like $x_i x_k x_m x_p \ldots$ is called a tensor of order n, where n is the number of factors in this product.

When we say that a certain set of quantities is a tensor of any order n, we mean that we have ways of specifying its components with respect to any set of axes, and that the components with regard to any two different sets of axes are related according to the rule appropriate to tensors of that order, and in particular to the products of the coordinates with n factors. For instance, if we say that u_i is a tensor of order 1, we are not simply defining u_j' as meaning $a_{ij}u_i$. We are supposing both that u_j' has a meaning, such as a displacement or a velocity, with reference to the axes of x_j', and that the value of each component is equal to $a_{ij}u_i$. Thus the statement that any set of quantities is a tensor is not a mere convention, but a statement capable of test and therefore needing proof. In (7), for example, our data are that u_i and u_j' are the components of a vector with regard to two different sets of axes. We prove that the sets of quantities obtained by multiplying both by the same quantity are related according to the vector rule; and therefore the products are vectors.

If we interchange j and l in (11), we get

$$w_{lj}' = a_{il} a_{kj} w_{ik} \qquad (12).$$

But on the right i and k are "dummy suffixes"; that is, they are to be given all possible values and the results added. It is unimportant which of them we call i and which k; we may therefore interchange them and get

$$w_{lj}' = a_{kl} a_{ij} w_{ki} = a_{ij} a_{kl} w_{ki} \qquad (13).$$

Thus w_{ki} transforms according to the same rule as w_{ik} and therefore is another tensor of the second order. The importance of this is that if we know the set of quantities arranged

$$\begin{pmatrix} w_{11} & w_{12} & w_{13} \\ w_{21} & w_{22} & w_{23} \\ w_{31} & w_{32} & w_{33} \end{pmatrix} \qquad (14)$$

to be a tensor of the second order, then the arrangement

$$\begin{pmatrix} w_{11} & w_{21} & w_{31} \\ w_{12} & w_{22} & w_{32} \\ w_{13} & w_{23} & w_{33} \end{pmatrix} \qquad (15)$$

is another tensor of the second order. Therefore the sets $(w_{ik} + w_{ki})$ and $(w_{ik} - w_{ki})$ are tensors of the second order. The first of these has the property that it is unaltered by interchanging i and k, and is therefore called a *symmetrical* tensor. The second has all its components reversed in sign when i and k are interchanged, and is called an *antisymmetrical* tensor. Clearly in an antisymmetrical tensor the "leading diagonal" components, i.e. those with i and k equal, are all zero. Also, since

$$w_{ik} = \tfrac{1}{2}(w_{ik} + w_{ki}) + \tfrac{1}{2}(w_{ik} - w_{ki}) \qquad (16),$$

we can consider any tensor of the second order as the sum of symmetrical and antisymmetrical parts.

The gradient of a scalar is a vector. For if U is a scalar, its gradient is $\partial U/\partial x_i$ or $\partial U/\partial x_j'$ according to the set of axes. But

$$\frac{\partial U}{\partial x_j'} = \frac{\partial x_i}{\partial x_j'}\frac{\partial U}{\partial x_i} = a_{ij}\frac{\partial U}{\partial x_i} \qquad (17),$$

so that the gradients transform according to the vector rule. Similarly the gradient of a vector is a tensor of order 2. For if u_i and u_j' are the components of a vector with respect to two sets of axes,

$$\frac{\partial u_j'}{\partial x_l'} = \frac{\partial x_k}{\partial x_l'}\frac{\partial u_j'}{\partial x_k} = a_{kl}\frac{\partial}{\partial x_k}(a_{ij}\,u_i)$$

$$= a_{ij}a_{kl}\frac{\partial u_i}{\partial x_k} \qquad (18),$$

so that the rule of transformation is as in (11).

Since x_i is a vector, it follows that $\partial x_i/\partial x_k$ is a tensor of the second order. But $\partial x_i/\partial x_k$ is unity if $i = k$ and zero if $i \neq k$. Hence the set of quantities δ_{ik}, such that

$$\delta_{11} = \delta_{22} = \delta_{33} = 1,$$

$$\delta_{12} = \delta_{13} = \delta_{21} = \delta_{23} = \delta_{31} = \delta_{32} = 0,$$

constitutes a tensor of the second order. We can prove this directly; for if we apply (11), δ_{jl}' in the new system of coordinates should be given by

$$\delta_{jl}' = a_{ij}a_{kl}\delta_{ik} \qquad (19).$$

The suffix k has to take all values 1, 2, 3. But if $k \neq i$, δ_{ik} is 0, and the corresponding term is zero. If $k = i$, $\delta_{ik} = 1$, and the result of the summation with regard to k is

$$\delta_{jl}' = a_{ij}a_{il} \qquad (20).$$

But the a_{ij} are the direction cosines of the axis of x_j' with regard to the x_i, and the a_{il} are those of x_l' with regard to x_i. Hence $a_{ij}a_{il}$ is the cosine of the angle between x_j' and x_l', and is equal to 1 if the axes are identical and to 0 if they are perpendicular. It follows that the result of the transformation is that $\delta_{jl}' = 1$ if $j = l$, and $\delta_{jl}' = 0$ if $j \neq l$. It follows that the set of quantities

$$\begin{pmatrix} 1 & 0 & 0 \\ 0 & 1 & 0 \\ 0 & 0 & 1 \end{pmatrix} \tag{21}$$

is transformed into itself by the rule (11) and therefore is a tensor of the second order.

If u_i is a vector and we form the product $\delta_{ik} u_m$, we have a tensor of order 3. But now put $m = k$ and add for all values of k. Since $\delta_{ik} = 0$ except for $k = i$, the only term different from zero is that for $k = i$, and this is u_i. Hence

$$\delta_{ik} u_k = u_i \qquad (22).$$

This operation therefore replaces the suffix k by i. The tensor δ_{ik} can therefore be called the *substitution tensor*.

In the tensor w_{ik} let us put $k = i$, and in accordance with our convention add for all values of i. Then the corresponding quantity w_{jj}' is got by putting $l = j$ and summing; but

$$\begin{aligned} w_{jj}' = a_{ij} a_{kj} w_{ik} &= \delta_{ik} w_{ik} \\ &= w_{ii} \end{aligned} \qquad (23).$$

Thus w_{ii} transforms into itself and therefore is a scalar.

This operation of putting two suffixes in a tensor equal and adding accordingly is known as *contraction*. In general it gives a new tensor, whose order is less by 2 than that of the original tensor. If for instance we contract the tensor $u_i v_k$, we obtain

$$u_i v_i = u_1 v_1 + u_2 v_2 + u_3 v_3 \qquad (24),$$

which is the *scalar product* of u_i and v_k.

Similarly the tensor $u_i v_k$ yields the symmetrical and antisymmetrical tensors $(u_i v_k + u_k v_i)$ and $(u_i v_k - u_k v_i)$. We may call these the *symmetrical* and *antisymmetrical* products of u_i and v_k.

The tensor $\partial u_k/\partial x_i$ gives similarly, on contraction, a scalar

$$\frac{\partial u_i}{\partial x_i} = \frac{\partial u_1}{\partial x_1} + \frac{\partial u_2}{\partial x_2} + \frac{\partial u_3}{\partial x_3} \qquad (25),$$

which is known as the *divergence* of u_i; while it gives also symmetrical and antisymmetrical tensors

$$\frac{\partial u_k}{\partial x_i} + \frac{\partial u_i}{\partial x_k} \quad \text{and} \quad \frac{\partial u_k}{\partial x_i} - \frac{\partial u_i}{\partial x_k}.$$

The former has important applications, especially in the theory of elasticity and hydrodynamics; the latter is known as the *curl* or *rotation* of u_i. The vanishing of the curl is the condition that u_i may be the gradient of a scalar.

All the above considerations can be extended to any number of dimensions. In n dimensions a tensor of order r has n^r components. A tensor of order 2, in particular, has n^2 components. If it is antisymmetrical, the n diagonal components are zero, and the others are equal and opposite in pairs. Hence an antisymmetrical tensor of order 2 has $\frac{1}{2}n\,(n-1)$ independent components. If $n = 1, 2, 3, 4,\ldots$ in turn, this number is $0, 1, 3, 6, \ldots$. It happens that in three dimensions the number of numerically independent components of an antisymmetrical tensor of the second order is equal to the number of components of a vector. Actually it can be proved that with any vector we can associate an antisymmetrical tensor of the second order, and conversely. This is not true in any number of dimensions other than 3.

Since the a_{ij} are the direction cosines with respect to the x_i of three perpendicular lines, they are connected by six relations

$$\left.\begin{aligned} a_{11}{}^2 + a_{21}{}^2 + a_{31}{}^2 &= 1 \\ a_{12}{}^2 + a_{22}{}^2 + a_{32}{}^2 &= 1 \\ a_{13}{}^2 + a_{23}{}^2 + a_{33}{}^2 &= 1 \end{aligned}\right\} \qquad (26),$$

$$\left.\begin{aligned} a_{12}a_{13} + a_{22}a_{23} + a_{32}a_{33} &= 0 \\ a_{13}a_{11} + a_{23}a_{21} + a_{33}a_{31} &= 0 \\ a_{11}a_{12} + a_{21}a_{22} + a_{31}a_{32} &= 0 \end{aligned}\right\} \qquad (27).$$

We notice that the second and third of (27) both contain (a_{11}, a_{21}, a_{31}). We may therefore solve them for the ratios of these quantities. Thus

$$\frac{a_{11}}{a_{23}a_{32}-a_{33}a_{22}}=\frac{a_{21}}{a_{33}a_{12}-a_{13}a_{32}}=\frac{a_{31}}{a_{13}a_{22}-a_{23}a_{12}}=k \quad (28),$$

say. Substituting in the first of (26) we get

$$\begin{aligned}1 &= k\{a_{11}(a_{23}a_{32}-a_{33}a_{22})+a_{21}(a_{33}a_{12}-a_{13}a_{32}) \\ &\qquad + a_{31}(a_{13}a_{22}-a_{23}a_{12})\} \\ &= -k\begin{vmatrix} a_{11} & a_{21} & a_{31} \\ a_{12} & a_{22} & a_{32} \\ a_{13} & a_{23} & a_{33}\end{vmatrix} \quad (29).\end{aligned}$$

Also

$$k^2\{(a_{23}a_{32}-a_{33}a_{22})^2+(a_{33}a_{12}-a_{13}a_{32})^2+(a_{13}a_{22}-a_{23}a_{12})^2\}=1 \quad (30).$$

But we have a general identity

$$\begin{aligned}(a^2+b^2+c^2)(a'^2+b'^2+c'^2)-(aa'+bb'+cc')^2 \\ = (bc'-cb')^2+(ca'-ac')^2+(ab'-ba')^2 \quad (31).\end{aligned}$$

Hence

$$\begin{aligned}k^2[(a_{13}{}^2+a_{23}{}^2+a_{33}{}^2)(a_{12}{}^2+a_{22}{}^2+a_{32}{}^2) \\ -(a_{12}a_{13}+a_{22}a_{23}+a_{32}a_{33})^2]=1 \quad (32).\end{aligned}$$

But on account of the second and third of (26) and the first of (27) the expression in brackets is unity, and therefore

$$k = \pm 1 \quad (33).$$

For any given transformation the determinant in (29) is therefore equal to ± 1. Evidently its sign is reversed if we interchange any two of the suffixes j, for this interchanges two rows of the determinant; so that the sign is a matter of the numbering of the axes. If we start with a rigid frame attached to the axes x_i, and rotate it continuously till it is attached to the axes x_j', all the a_{ij} vary continuously and therefore the determinant cannot change from $+1$ to -1

or from -1 to $+1$. If then x_1 goes to x_1', x_2 to x_2', and x_3 to x_3', the determinant is initially

$$\begin{vmatrix} 1 & 0 & 0 \\ 0 & 1 & 0 \\ 0 & 0 & 1 \end{vmatrix} = 1 \qquad (34),$$

and therefore

$$k = -1 \qquad (35),$$

and the determinant formed by the a_{ij} is always $+1$.

If we have a frame of axes $(x_1 x_2 x_3)$ we can turn it by a continuous movement so as to bring x_1 along the old x_2, x_2 along the old x_3, and x_3 along the old x_1. In this case we have

$$x_1' = x_2, \quad x_2' = x_3, \quad x_3' = x_1 \qquad (36),$$

and

$$\left.\begin{array}{c} a_{11} = 0,\ a_{21} = 1,\ a_{31} = 0;\ a_{12} = 0,\ a_{22} = 0,\ a_{32} = 1; \\ a_{13} = 1,\ a_{23} = 0,\ a_{33} = 0 \end{array}\right\} \qquad (37).$$

The determinant of the a_{ij} is therefore

$$\begin{vmatrix} 0 & 1 & 0 \\ 0 & 0 & 1 \\ 1 & 0 & 0 \end{vmatrix} = 1 \qquad (38),$$

as before. Any rotation of the axes that does not alter the cyclic interchange of suffixes 1 2 3 1 2 3 1 ... therefore leaves the determinant equal to unity, and therefore so long as we always use right-handed or always left-handed axes the determinant of the a_{ij} is $+1$.

With this restriction

$$\left.\begin{array}{c} a_{11} = a_{22}a_{33} - a_{32}a_{23};\ a_{21} = a_{13}a_{32} - a_{33}a_{12}; \\ a_{31} = a_{23}a_{12} - a_{13}a_{22} \end{array}\right\} \qquad (39),$$

and therefore every direction cosine is equal to its first minor in the determinant.

These relations are of course identical with those expressed in the usual notation of solid geometry by

$$l_1 = m_2 n_3 - m_3 n_2;\ m_1 = n_2 l_3 - n_3 l_2;\ n_1 = l_2 m_3 - l_3 m_2 \qquad (40).$$

Now suppose that u_i is a vector, and consider the set of quantities

$$w_{ik} = \begin{pmatrix} 0 & u_3 & -u_2 \\ -u_3 & 0 & u_1 \\ u_2 & -u_1 & 0 \end{pmatrix} \qquad (41).$$

Apply (11) to this, taking i to be the number of the row and k that of the column. We see that u_1 enters as w_{23} and as $-w_{32}$. Its coefficient in w_{jl}' is therefore $a_{2j}a_{3l} - a_{3j}a_{2l}$, and in all

$$w_{jl}' = (a_{2j}a_{3l} - a_{3j}a_{2l})\,u_1 + (a_{3j}a_{1l} - a_{1j}a_{3l})\,u_2 + (a_{1j}a_{2l} - a_{2j}a_{1l})\,u_3 \qquad (42).$$

This is obviously zero if $j = l$. If $j \neq l$ and if the other axis perpendicular to x_j' and x_l' is x_n', and $jlnjln$ is a cyclic order, the quantities in brackets are equal to (a_{1n}, a_{2n}, a_{3n}). This is true if l immediately succeeds j in the order. If l precedes j by one place the signs are reversed. Hence if $j = 1$ and $l = 2$, or $j = 2$ and $l = 3$, or if $j = 3$ and $l = 1$,

$$w_{jl}' = a_{in}u_i = u_n' \qquad (43),$$

and in the alternative case

$$w_{jl}' = -u_n' \qquad (44).$$

Thus
$$w_{jl}' = \begin{pmatrix} 0 & u_3' & -u_2' \\ -u_3' & 0 & u_1' \\ u_2' & -u_1' & 0 \end{pmatrix} \qquad (45),$$

and is of the same form as (41). Thus with any vector we can associate an antisymmetrical tensor of the second order. Conversely with any antisymmetrical tensor of the second order we can associate a vector.

We can proceed alternatively by considering the set of quantities ϵ_{ikm}, defined by the condition that if any two of i, k, m are equal the corresponding component is 0; if i, k, m are all unequal and in cyclic order, the component is $+1$; if the order is not cyclic, the component is -1. Let us see whether this is a tensor of the third order. If so, we should have

$$\begin{aligned}\epsilon_{jln}' &= a_{ij}a_{kl}a_{mn}\epsilon_{ikm}\\ &= a_{1j}a_{2l}a_{3n} + a_{2j}a_{3l}a_{1n} + a_{3j}a_{1l}a_{2n}\\ &\quad - a_{2j}a_{1l}a_{3n} - a_{3j}a_{2l}a_{1n} - a_{1j}a_{3l}a_{2n} \qquad (46).\end{aligned}$$

Now if, for instance, $j = l$, the right side is clearly zero and $\epsilon_{jln}' = 0$. If j, l, n are all unequal, the expression is

$$\begin{vmatrix} a_{1j} & a_{2j} & a_{3j} \\ a_{1l} & a_{2l} & a_{3l} \\ a_{1n} & a_{2n} & a_{3n} \end{vmatrix} \qquad (47),$$

which is equal to 1 if jln are in cyclic order and to -1 if not. Hence the set of quantities ϵ_{ikm} is transformed into itself by the rule for transforming tensors of order 3, and therefore constitutes a tensor of order 3. This is called the *alternating tensor*.

Now consider the product $\epsilon_{ikm}u_p$, where u_p is a vector. This is a tensor of the fourth order. If we contract it by putting $p = m$ and summing we get a second order tensor $w_{ik} = \epsilon_{ikm}u_m$. If $i = 1$ and $k = 2$, the only value of m that makes ϵ_{12m} different from zero is 3, and then $\epsilon_{ikm} = +1$. Hence

$$w_{12} = u_3 \qquad (48).$$

If $i = 2$ and $k = 1$, m is 3; but 213 is the reverse of cyclic order and $\epsilon_{213} = -1$. Hence

$$w_{21} = -u_3 \qquad (49).$$

Similarly we find that the elements of w_{ik} are

$$\begin{pmatrix} 0 & u_3 & -u_2 \\ -u_3 & 0 & u_1 \\ u_2 & -u_1 & 0 \end{pmatrix} \qquad (50),$$

so that the antisymmetrical tensor associated with a vector can actually be obtained from it by multiplying by ϵ_{ikm} and contracting.

Again, suppose that we are given a tensor of the second order w_{ik} and that we form a vector u_m by multiplying by ϵ_{ikm} and contracting twice. We have, if $m = 3$,

$$u_3 = \epsilon_{ik3} w_{ik} = \epsilon_{123} w_{12} + \epsilon_{213} w_{21} = w_{12} - w_{21}.$$

Thus
$$u_m = \epsilon_{ikm} w_{ik} \qquad (51).$$

If w_{ik} is symmetrical, this evidently gives zero. If it is antisymmetrical the components of u_m are numerically twice those of w_{ik}.

On account of the intimate relation between the vector and the antisymmetrical tensor we shall habitually denote the tensor w_{ik} of (41) by u_{ik}, so that

$$\left.\begin{aligned} u_{11} = u_{22} = u_{33} = 0;\ u_{12} = u_3,\ u_{23} = u_1,\ u_{31} = u_2; \\ u_{21} = -u_3,\ u_{32} = -u_1,\ u_{13} = -u_2 \end{aligned}\right\} \qquad (52).$$

It will always be seen at once whether the vector or the tensor is intended, since the former has one and the latter two suffixes.

If we have any three vectors u_i, v_i, w_i, and consider the scalar $\epsilon_{ikm} u_i v_k w_m$, we see that

$$\begin{aligned} \epsilon_{ikm} u_i v_k w_m &= u_1 v_2 w_3 + u_2 v_3 w_1 + u_3 v_1 w_2 \\ &\qquad - u_2 v_1 w_3 - u_3 v_2 w_1 - u_1 v_3 w_2 \\ &= \begin{vmatrix} u_1 & u_2 & u_3 \\ v_1 & v_2 & v_3 \\ w_1 & w_2 & w_3 \end{vmatrix} \end{aligned} \qquad (53),$$

so that we have a concise way of writing the determinant formed by the components of three vectors. If any two of the vectors are parallel this scalar vanishes.

In associating a vector with an antisymmetrical tensor of order 2 a sign convention clearly arises. We make the positive signs in (41) lie one place to the *right* of the leading diagonal. If then we have two vectors u_i and v_k, their antisymmetrical product is $u_i v_k - u_k v_i$, and in the associated vector we give the positive sign to w_{ik} when k follows i in the cyclic order. Hence the components of this vector are taken to be

$$(u_2 v_3 - u_3 v_2,\ u_3 v_1 - u_1 v_3,\ u_1 v_2 - u_2 v_1).$$

This vector is perpendicular to both the original vectors; for

$$u_1 (u_2 v_3 - u_3 v_2) + u_2 (u_3 v_1 - u_1 v_3) + u_3 (u_1 v_2 - u_2 v_1) = 0,$$

$$v_1 (u_2 v_3 - u_3 v_2) + v_2 (u_3 v_1 - u_1 v_3) + v_3 (u_1 v_2 - u_2 v_1) = 0.$$

We call it the *vector product* of u_i and v_k, and can save writing by denoting it by $[u, v]_m$.

Similarly with the antisymmetrical tensor $\frac{\partial u_k}{\partial x_i} - \frac{\partial u_i}{\partial x_k}$ we associate a vector so as to leave the sign unaltered when k follows i in the cyclic order. Thus the components are

$$\left(\frac{\partial u_3}{\partial x_2} - \frac{\partial u_2}{\partial x_3},\ \frac{\partial u_1}{\partial x_3} - \frac{\partial u_3}{\partial x_1},\ \frac{\partial u_2}{\partial x_1} - \frac{\partial u_1}{\partial x_2}\right) \qquad (54).$$

This is often called "curl u."

The Tensor $\epsilon_{iks}\epsilon_{mps}$. Since this tensor is the product of two third order tensors, once summed, it is a tensor of the fourth order, i, k, m, p being arbitrarily assignable. Evidently if $i = k$ or $m = p$, the corresponding component is zero.

If $i = m$, the contribution from any value of s is zero unless also $k = p$, and then

$$\epsilon_{iks} = \epsilon_{mps} = \pm 1,$$

and the component is $+ 1$.

If $i = p$, then no value of s gives a contribution unless $k = m$. Then one of ϵ_{iks} and ϵ_{mps} is $+ 1$ and the other $- 1$, and the component is $- 1$. Hence the components of the tensor are as follows.

If $i = m, k = p$, the component is $+ 1$, unless i also $= k$,
$i = p, k = m$, the component is $- 1$, unless i also $= k$,
$i = k$ or $m = p$, the component is 0.

These results apply also to the tensor

$$\delta_{im}\delta_{kp} - \delta_{ip}\delta_{km},$$

and therefore

$$\epsilon_{iks}\epsilon_{mps} = \delta_{im}\delta_{kp} - \delta_{ip}\delta_{km} \quad (55).$$

EXAMPLES

1. If u_i, u_j', u_k'' are the components of a vector with regard to three sets of axes, prove that the values of u_k'' are the same as would be obtained by transforming first from u_i to u_j' and then from u_j' to u_k''.

2. Prove that $\delta_{ii} = 3;\ \delta_{ik}\epsilon_{ikm} = 0.$

3. Evaluate the components of the sixth-order tensor

$$\epsilon_{abc}\epsilon_{ikm}.$$

4. Prove that

$$\epsilon_{iks}\epsilon_{mks} = 2\delta_{im};\quad \epsilon_{ikm}\epsilon_{ikm} = 6.$$

5. Prove that

$$\epsilon_{iks}\epsilon_{mps} = \epsilon_{sik}\epsilon_{smp} = \epsilon_{ksi}\epsilon_{psm}.$$

6. Prove that if u_i, v_k, w_m are vectors,

$$[w\,[u, v]]_m = u_m\,(v_i w_i) - v_m\,(u_i w_i).$$

$$u_m\,[v, w]_m = \epsilon_{ikm} u_i v_k w_m,$$

7. If

$$\Delta(u) = \begin{vmatrix} u_{11} & u_{12} & u_{13} \\ u_{21} & u_{22} & u_{23} \\ u_{31} & u_{32} & u_{33} \end{vmatrix},$$

prove that

$$\epsilon_{ijk}\,\Delta(u) = \epsilon_{lmn} u_{il} u_{jm} u_{kn},$$

$$\epsilon_{ijk} u_{il} u_{jm} u_{kn} = \epsilon_{lmn}\,\Delta(u),$$

$$6\Delta(u) = \epsilon_{ijk}\epsilon_{lmn} u_{il} u_{jm} u_{kn}.$$

8. Use Ex. **7** to prove the rule for the multiplication of determinants

$$\epsilon_{lmn}\epsilon_{lmn}\,\Delta(u)\,\Delta(v) = 6\Delta(u)\,\Delta(v) = 6\Delta(uv),$$

where

$$(uv)_{ip} = u_{il} v_{lp}.$$

CHAPTER II

GEOMETRICAL APPLICATIONS

The displacement from any point to any other obviously constitutes a vector. The distance between the points is a scalar. If x_i, y_i are the coordinates of the points and r the distance between them,

$$r^2 = (y_i - x_i)^2 \qquad (1),$$

the square on the right indicating the scalar product of the vector into itself. Also the quantities $(y_i - x_i)/r$ constitute a vector.

If we take a fixed point a_i and consider points given by

$$x_i = a_i + l_i r \qquad (2),$$

where r is a variable scalar and the l_i are constants such that

$$l_i^2 = 1 \qquad (3),$$

$$(x_i - a_i)^2 = r^2 \qquad (4),$$

so that r is the distance of x_i from a_i. If we take another point y_i such that

$$y_i = a_i + l_i s \qquad (5),$$

$$(y_i - a_i)^2 = s^2 \qquad (6),$$

$$(y_i - x_i)^2 = (s - r)^2 \qquad (7),$$

and therefore the distances between a_i, x_i, and y_i are such that the sum of two of them is equal to the third. Thus the points are on a straight line; and (2) gives the equations of the line in terms of the parameter r. The l_i are the *direction cosines* of the line.

If we take two lines through a_i given by

$$x_i = a_i + l_i r \qquad (8),$$

$$y_i = a_i + m_i s \qquad (9),$$

the distance between x_i and y_i is given by

$$(y_i - x_i)^2 = (sm_i - rl_i)^2$$
$$= s^2 + r^2 - 2rsl_im_i \qquad (10).$$

But this quantity is also equal to $s^2 + r^2 - 2rs \cos \theta$, where θ is the angle between the lines. Hence the angle between two intersecting lines is given by

$$\cos \theta = l_im_i \qquad (11).$$

If two lines have the same direction cosines they are said to be *parallel*. If two lines do not intersect we can take a line through any point on one of them parallel to the other; then this line is inclined to the first at an angle given by (11). We can then use (11) to determine a unique quantity associated with any two lines, which we may call their inclination, whether they intersect or not.

If we have a line given by (2) and y_i is a point outside it, the line joining α_i and y_i subtends a right angle at x_i if

$$(y_i - \alpha_i)^2 = (x_i - \alpha_i)^2 + (y_i - x_i)^2$$
$$= r^2l_i^2 + (y_i - \alpha_i - rl_i)^2$$
$$= (y_i - \alpha_i)^2 - 2rl_i(y_i - \alpha_i) + 2r^2l_i^2 \quad (12),$$

and therefore

$$r = l_i(y_i - \alpha_i) \qquad (13).$$

This gives the projection of the displacement $y_i - \alpha_i$ on the line. The foot of the perpendicular is

$$\alpha_i + l_ir = \alpha_i + l_il_k(y_k - \alpha_k).$$

Evidently r in (13) will be the same for all points y_i such that l_iy_i is constant. Hence

$$l_iy_i = \delta \qquad (14)$$

represents a plane perpendicular to the line.

If we take two intersecting lines given by (8) and (9), we

can find the equation of the plane containing them as follows. If this plane is

$$n_i z_i = p \qquad (15),$$

this equation must be satisfied by x_i and y_i for all values of r and s. Hence

$$n_i \alpha_i = p \qquad (16),$$

$$n_i l_i = 0 \qquad (17),$$

$$n_i m_i = 0 \qquad (18),$$

and from (15) and (16),

$$n_i (z_i - \alpha_i) = 0 \qquad (19).$$

Then (17), (18), (19) are three homogeneous equations in the n_i, and can be consistent only if

$$\epsilon_{ikm} (z_i - \alpha_i) l_k m_m = 0 \qquad (20).$$

This is the equation of the required plane. Also the n_i are proportional to

$$\epsilon_{ikm} l_k m_m = (l_2 m_3 - l_3 m_2, \ l_3 m_1 - l_1 m_3, \ l_1 m_2 - l_2 m_1) \quad (21).$$

But

$$n_i^2 = 1 \qquad (22),$$

$$\begin{aligned}(l_2 m_3 - l_3 m_2)^2 &+ (l_3 m_1 - l_1 m_3)^2 + (l_1 m_2 - l_2 m_1)^2 \\ &= (l_1^2 + l_2^2 + l_3^2)(m_1^2 + m_2^2 + m_3^2) - (l_1 m_1 + l_2 m_2 + l_3 m_3)^2 \\ &= 1 - \cos^2 \theta \\ &= \sin^2 \theta \qquad (23).\end{aligned}$$

Thus

$$\sin \theta \, n_i = \pm \epsilon_{ikm} l_k m_m \qquad (24).$$

The ambiguity in sign corresponds to a general one in specifying the parameter r of a point on a line. If r in (8) is taken negative, we get a point on the line on the opposite side of α_i from those given by positive values of r. But if we reverse both r and the l_i we still keep $l_i^2 = 1$, and we still have the same point. We may take either direction along a line to be that of r increasing; if we reverse the direction the signs of all the l_i are reversed for the same point.

For any point on (8),

$$\epsilon_{ikm} x_k l_m = \epsilon_{ikm} (\alpha_k + l_k r) l_m = \epsilon_{ikm} \alpha_k l_m \qquad (25).$$

This is a constant vector for all values of r, and may therefore be considered as a property of the line. We denote it by l_i'. Then we have six properties of the line given by l_i, l_i'. These are coordinates of the line. They are connected by two relations,

$$l_i^2 = 1 \qquad (26),$$

$$l_i l_i' = \epsilon_{ikm} l_i \alpha_k l_m = 0 \qquad (27).$$

The l_i' have a geometrical interpretation. Thus if we consider the plane

$$x_2 l_3 - x_3 l_2 = l_1' \qquad (28),$$

this plane passes through the line. Also if $x_2 = \alpha_2$, $x_3 = \alpha_3$, (28) is satisfied for all values of x_1, and therefore if we take a line through α_i parallel to the x_1 axis, (28) represents the plane through (8) and this line. Two such planes determine the line, and therefore the l_i and l_i' together determine the line.

If we have two non-intersecting lines given by

$$x_i = \alpha_i + r l_i \qquad (29),$$

$$y_i = \beta_i + s m_i \qquad (30),$$

the line

$$y_i = \beta_i + r l_i \qquad (31)$$

passes through β_i and is parallel to (29). The plane including (30) and (31) is, by (20),

$$\epsilon_{ikm} (z_i - \beta_i) l_k m_m = 0 \qquad (32).$$

This therefore represents a plane through (30) parallel to (29). The plane through (29) parallel to (30) is

$$\epsilon_{ikm} (z_i - \alpha_i) l_k m_m = 0 \qquad (33).$$

The distance between these planes is the projection of the line joining any two points on them upon a line perpendicular to both. If a line perpendicular to both has direction cosines n_i, the shortest distance d between the lines is therefore given by

$$d \sin\theta = (\beta_i - \alpha_i)\, n_i \sin\theta$$
$$= \pm (\beta_i - \alpha_i)\, \epsilon_{ikm} l_k m_m$$
$$= \pm \{\epsilon_{ikm} \beta_i l_k m_m - \epsilon_{ikm} \alpha_i l_k m_m\} \qquad (34).$$

But
$$\epsilon_{ikm} \beta_i m_m = - m_k'; \; \epsilon_{ikm} \alpha_i l_k = l_m' \qquad (35),$$
and therefore

$$d \sin\theta = \pm (- l_i m_i' - m_i l_i'),$$

so that, apart from the ambiguity in sign,

$$d \sin\theta = l_i m_i' + m_i l_i' \qquad (36).$$

Thus the shortest distance is directly expressible in terms of the coordinates of the two lines.

Now consider two intersecting lines

$$x_i = \alpha_i + r l_i; \; y_i = \alpha_i + s m_i \qquad (37).$$

The area of the triangle formed by α_i, x_i, y_i is $\frac{1}{2} rs \sin\theta$. The projections of these points on the plane $x_1 = 0$ are $(0, \alpha_2, \alpha_3)$, $(0, x_2, x_3)$, $(0, y_2, y_3)$ and form a triangle whose area is

$$\frac{1}{2}\begin{vmatrix} 1 & \alpha_2 & \alpha_3 \\ 1 & x_2 & x_3 \\ 1 & y_2 & y_3 \end{vmatrix} = \frac{1}{2}\begin{vmatrix} 1 & \alpha_2 & \alpha_3 \\ 0 & l_2 r & l_3 r \\ 0 & m_2 s & m_3 s \end{vmatrix}$$
$$= \frac{1}{2} rs\, (l_2 m_3 - l_3 m_2)$$
$$= \frac{1}{2} rs n_1 \sin\theta \qquad (38).$$

Thus the projections of a triangle, and therefore of any plane area, on the coordinate planes are in the ratios of the direction cosines of the normal to the planes. A plane area can therefore be treated as a vector whose components are proportional to the direction cosines of the normal.

If we have a line given by

$$x_i = \alpha_i + rl_i \qquad (39),$$

and β_i is a point not on the line, let us suppose the point β_i turned through an angle θ about the line. The foot of the normal from β_i to the line is given by

$$r = l_i (\beta_i - \alpha_i) \qquad (40),$$

and therefore the displacement from the foot of the normal to β_i is equal to

$$\begin{aligned}\beta_i - \{\alpha_i + l_i l_k (\beta_k - \alpha_k)\} &= l_k^2 (\beta_i - \alpha_i) - l_i l_k (\beta_k - \alpha_k) \\ &= l_k \{l_k (\beta_i - \alpha_i) - l_i (\beta_k - \alpha_k)\} \qquad (41).\end{aligned}$$

The magnitude of this displacement, p, is given by

$$\begin{aligned}p^2 &= (\beta_i - \alpha_i)^2 - \{l_i (\beta_i - \alpha_i)\}^2 \\ &= l_k^2 (\beta_i - \alpha_i)^2 - l_i l_k (\beta_i - \alpha_i)(\beta_k - \alpha_k) \\ &= \tfrac{1}{2} \{l_k (\beta_i - \alpha_i) - l_i (\beta_k - \alpha_k)\}^2 \qquad (42),\end{aligned}$$

the $\frac{1}{2}$ being needed because in the double summation each pair of values of the suffixes would occur twice.

The plane through β_i and the line is

$$n_i (z_i - \alpha_i) = 0,$$

subject to

$$n_i l_i = 0,$$

$$n_i (\beta_i - \alpha_i) = 0,$$

and is therefore

$$\epsilon_{ikm} (z_i - \alpha_i) l_k (\beta_m - \alpha_m) = 0 \qquad (43),$$

while the n_i are proportional to $\epsilon_{ikm} l_k (\beta_m - \alpha_m)$ and therefore equal to $\pm \epsilon_{ikm} l_k (\beta_m - \alpha_m)/p$.

If now we turn β_i through an angle θ, it receives a displacement $p(1 - \cos\theta)$ along the normal to (39), and a displacement $p \sin\theta$ along the perpendicular to the plane (43). If it goes to γ_i, we have therefore

$$\begin{aligned}\gamma_i - \beta_i = &- (1 - \cos\theta) l_k \{l_k (\beta_i - \alpha_i) - l_i (\beta_k - \alpha_k)\} \\ &\pm \sin\theta \, \epsilon_{ikm} l_k (\beta_m - \alpha_m) \qquad (44).\end{aligned}$$

If θ is a small angle and we neglect θ^2, the displacement is simply

$$\gamma_i - \beta_i = \pm\, \theta\, \epsilon_{ikm} l_k (\beta_m - \alpha_m) \qquad (45),$$

and in particular

$$\gamma_1 - \beta_1 = \pm \{l_2 \theta (\beta_3 - \alpha_3) - l_3 \theta (\beta_2 - \alpha_2)\} \qquad (46),$$

and from the additive form of this equation we see that the displacement is the sum of those given by separate small rotations $l_i \theta$ about axes through α_i parallel to the coordinate axes. Conversely, displacements due to small rotations about axes through a point can be added vectorially as if all were applied to the system in its original position, and give the same total displacement as if they were compounded into a single rotation about an axis by the vector rule. We still have, however, to establish a sign convention. We decide that θ is to be taken positive if a turn about the axis of x_3 is from x_1 towards x_2. This would make

$$\gamma_1 - \beta_1 = -\theta (\beta_2 - \alpha_2);\; \gamma_2 - \beta_2 = \theta (\beta_1 - \alpha_1) \qquad (47)$$

with $l_1 = l_2 = 0$, $l_3 = 1$. Hence

$$\gamma_i - \beta_i = \epsilon_{ikm} l_k \theta (\beta_m - \alpha_m) \qquad (48).$$

If we write

$$l_k \theta = \varpi_k,$$

$$\gamma_i - \beta_i = \epsilon_{ikm} \varpi_k (\beta_m - \alpha_m) \qquad (49).$$

For instance, if $\alpha_m = 0$, we have

$$\gamma_1 - \beta_1 = \varpi_2 \beta_3 - \varpi_3 \beta_2;\; \gamma_2 - \beta_2 = \varpi_3 \beta_1 - \varpi_1 \beta_3;$$

$$\gamma_3 - \beta_3 = \varpi_1 \beta_2 - \varpi_2 \beta_1 \qquad (50).$$

For finite rotations we return to (44) or (41), keeping the positive sign in the second term. We may transfer the origin to α_i to save writing. The coefficient of β_k in $\gamma_i - \beta_i$ is

$$b_{ik} = (1 - \cos\theta)\, l_i l_k - \sin\theta\, \epsilon_{ikm} l_m \qquad (51),$$

for $k \neq i$; if $k = i$, the coefficient is $-(1 - \cos\theta)(1 - l_i l_k)$. Thus

$$b_{ik} = (1 - \cos\theta)(l_i l_k - \delta_{ik}) - \sin\theta\, c_{ik} \qquad (52),$$

where c_{ik} is the antisymmetrical tensor corresponding to l_m, namely

$$\begin{pmatrix} 0 & l_3 & -l_2 \\ -l_3 & 0 & l_1 \\ l_2 & -l_1 & 0 \end{pmatrix} \qquad (53).$$

Thus the displacement is represented in general by $b_{ik}\beta_k$, where b_{ik} is a tensor of the second order, expressed as the sum of symmetrical and antisymmetrical tensors. The antisymmetrical part is seen to be of the first order of magnitude in θ and the symmetrical part of the second order.

EXAMPLES

1. Given that the general quadric surface is

$$S \equiv \tfrac{1}{2}A_{ik}x_i x_k + B_i x_i + C = 0,$$

prove that the locus of the mid-points of parallel chords is a plane, and find the condition for this plane to be perpendicular to the chords.

2. Find the condition that the line

$$l_i x_i + p = 0$$

may touch the quadric S.

CHAPTER III

PARTICLE DYNAMICS

The essence of particle dynamics is that the second derivatives of the coordinates of a particle with regard to the time are equal to functions of the position and velocity of the particle with reference to neighbouring particles. The relations therefore provide a set of differential equations to determine the coordinates. The equations for any particle can be put in the form

$$m\ddot{x}_i = \Sigma X_i \qquad (1),$$

where m is the mass of the particle, X_i are the forces due to the other particles, and the summation is for all the other particles*. It is a matter of experiment that this form holds when the axes are a certain type of Cartesian axes, which we call non-rotating and unaccelerated, or in brief *dynamical*.

If we take a different set of non-rotating axes at the same origin we have

$$x_j' = a_{ij} x_i \qquad (2),$$

and since the axes are not rotating the a_{ij} are constants. Hence by differentiation

$$\dot{x}_j' = a_{ij}\dot{x}_i \qquad (3),$$

$$\ddot{x}_j' = a_{ij}\ddot{x}_i \qquad (4),$$

and therefore velocity and acceleration are vectors.

The force X_i on the particle due to some other particle is measured by the contribution to $m\ddot{x}_i$ due to the other particle; that is, the part of $m\ddot{x}_i$ that would disappear if the other particle was removed to an infinite distance. It

* Cf. Jeffreys, *Scientific Inference*, Chapter VIII, for a fuller analysis.

is actually measured by the acceleration. If we have the three acceleration components of the particle 1 due to the particle 2, we can find the acceleration in the direction of x_j' by the formula (4), and the form (1) will become

$$m\ddot{x}_j' = \Sigma X_j' \qquad (5),$$

provided that we define X_j' by the rule

$$X_j' = a_{ij} X_i \qquad (6).$$

The meaning of the force in any direction requires definition in any case: if we define it by its relation to the acceleration component in that direction, it follows automatically by (6) that force is a vector. It follows that the sum of any number of forces, obtained by adding the components separately, is also a vector. But the practical importance of the idea of force arises equally from the fact that in many cases the force components are known once for all from experience as functions of the coordinates and velocities. The form (6) then says, as a general principle, that the forces in any direction are additive.

If we form the contracted or scalar product of $m\ddot{x}_i$ and ΣX_i by the vector $\dot{x}_i$, we get

$$m\dot{x}_i \ddot{x}_i = \Sigma X_i \dot{x}_i \qquad (7),$$

the left side of which is $\dfrac{d}{dt}(\tfrac{1}{2}m\dot{x}_i^2)$. (The square of course implies the product of $\dot{x}_i$ by $\dot{x}_i$ and therefore the summation for $i = 1, 2, 3$.) We write

$$T = \tfrac{1}{2}m\dot{x}_i^2 = \tfrac{1}{2}m(\dot{x}_1^2 + \dot{x}_2^2 + \dot{x}_3^2) \qquad (8),$$

and call T the *kinetic energy* of the particle. Then by integration with regard to the time from t_0 to t_1 we get

$$\Big[T\Big]_{t_0}^{t_1} = \Sigma \int_{t_0}^{t_1} X_i \dot{x}_i \, dt \qquad (9).$$

But in any short interval of time dt, $\dot{x}_i dt$ is the increase of x_i, namely dx_i. Hence the right side is equal to

$\Sigma \int X_i dx_i$ taken from the initial to the final position of the particle. We call the scalar $X_i dx_i$ the *work done* on the particle by the force X_i in a small displacement dx_i. Then $\Sigma \int X_i dx_i$ is the total work done by all the forces on the particle during the motion. We have therefore the scalar relation that in any motion of a particle

$$\text{Increase of kinetic energy} = \text{Work done on the particle} \qquad (10).$$

In a system of particles we may add up the equations (10) for all the particles. We now take T, the kinetic energy of the whole system, as the sum of those of all the particles. Then we get

$$\Big[T\Big]_{t_0}^{t_1} = \Big[\Sigma \tfrac{1}{2} m \dot{x}_i^2\Big]_{t_0}^{t_1} = \Sigma\Sigma \int_{t_0}^{t_1} X_i dx_i \qquad (11),$$

the double summation implying summation for all pairs of particles; the first summation is for the particles producing the forces X_i and the second for the particles acted on. It may happen that the X_i are all functions of the coordinates alone, and not of their velocities, and that provided the initial and final values of the x_i are the same the integral is the same however the x_i may vary in the interval. If so, the integral is the difference between the values of a certain function U for the initial and final positions of the system; we call the system *conservative* and U the *work-function*. Then (11) becomes

$$\Big[T\Big]_{t_0}^{t_1} = \Big[U\Big]_{t_0}^{t_1} \qquad (12),$$

or

$$T - U = \text{constant} \qquad (13).$$

This is the equation of conservation of energy. The quantities on both sides are scalars. $-U$ is often denoted by V and called the *potential energy*.

If U exists, then in all possible small displacements of the system

$$\Sigma\Sigma X_i dx_i = dU \qquad (14),$$

where the dx_i, $3n$ in number for n particles, are all independent. The coefficient of dx_i for any one particle is ΣX_i, the total force in the direction of x_i on that particle. Hence the force on a particle in the direction of x_i is $\partial U/\partial x_i$, which is of course a vector.

We can also take the scalar product of (1) by any set of small quantities δx_i whatever that constitute a vector. Then

$$m\ddot{x}_i \delta x_i = \Sigma X_i \delta x_i \qquad (15)$$

is a scalar relation; but as the δx_i are arbitrary we can equate their coefficients and regenerate the equation (1). If we now integrate with regard to t from t_0 to t_1 we get

$$\int_{t_0}^{t_1} m\ddot{x}_i \delta x_i\, dt = \int_{t_0}^{t_1} \Sigma X_i \delta x_i\, dt \qquad (16).$$

The left side, on integration by parts, gives

$$\Big[m\dot{x}_i \delta x_i\Big]_{t_0}^{t_1} - \int_{t_0}^{t_1} m\dot{x}_i \frac{d}{dt}\delta x_i\, dt \qquad (17).$$

But we can consider the $x_i + \delta x_i$ as coordinates of a particle in a motion differing slightly from the actual one; that is, at a given value of t, the coordinates are $x_i + \delta x_i$ instead of x_i. Then

$$\begin{aligned} \frac{d}{dt}\delta x_i &= \frac{d}{dt}(x_i + \delta x_i) - \frac{d}{dt}x_i \\ &= \dot{x}_i + \delta\dot{x}_i - \dot{x}_i \\ &= \delta\dot{x}_i \qquad (18), \end{aligned}$$

since $\dot{x}_i + \delta\dot{x}_i$ is simply the varied velocity or rate of change of the varied coordinate $x_i + \delta x_i$. Then

$$\begin{aligned} \int_{t_0}^{t_1} m\dot{x}_i \frac{d}{dt}\delta x_i\, dt &= \int_{t_0}^{t_1} m\dot{x}_i \delta\dot{x}_i\, dt \\ &= \int_{t_0}^{t_1} \delta\left(\tfrac{1}{2}m\dot{x}_i^2\right) dt + O(\delta\dot{x}_i)^2 \qquad (19). \end{aligned}$$

Then for every particle, to the first order,

$$\int_{t_0}^{t_1} \{\delta\,(\tfrac{1}{2}m\dot{x}_i^{\,2}) + \Sigma X_i\,\delta x_i\}\,dt = \Big[m\dot{x}_i\,\delta x_i\Big]_{t_0}^{t_1} \qquad (20).$$

We can now add up these equations for all the particles of the system. If U exists we can express the result in the form

$$\int_{t_0}^{t_1} \delta\,(T + U)\,dt = \Sigma\,\Big[m\dot{x}_i\,\delta x_i\Big]_{t_0}^{t_1} \qquad (21),$$

or, since the limits are not varied, we can move the δ outside the integral. If the variation is such that the initial and final positions of the system are unaltered, $\delta x_i = 0$ when $t = t_0$ or t_1, and we have

$$\delta \int_{t_0}^{t_1} (T + U)\,dt = 0$$

to the first order in the variations of the path. This is Hamilton's principle.

EXAMPLES

1. If the Cartesian coordinates of every particle of a system are known functions of a set of generalized coordinates q_r, prove that

$$\Sigma m\ddot{x}_i\,\delta x_i = \left\{\frac{d}{dt}\left(\frac{\partial T}{\partial \dot{q}_r}\right) - \frac{\partial T}{\partial q_r}\right\}\delta q_r,$$

$$\Sigma X_i\,\delta x_i = \Sigma X_i \frac{\partial x_i}{\partial q_r}\,\delta q_r,$$

where the summation convention is also understood on the right side. Deduce Lagrange's equations for the case where the q_r are all independent.

2. The equations of motion of a particle are $m\ddot{x}_i = X_i - k\dot{x}_i$ where k is constant. Prove that

$$2T = -\,x_i X_i + \frac{d^2}{dt^2}(\tfrac{1}{2}m x_i^{\,2}) + \frac{d}{dt}(\tfrac{1}{2}k x_i^{\,2}).$$

Hence show that for a system in periodic motion, or in one slowly changing its state, on an average over a long time, $2T = -\,\Sigma x_i X_i$.

CHAPTER IV

DYNAMICS OF RIGID BODIES

A rigid body is one such that whenever it is displaced the distance between any two particles of it is unaltered. Since three particles A, B, C are in a straight line if the sum of two of the distances AB, BC, CA is equal to the third, it follows that straight lines are unaltered by displacements of a rigid body. Since when A, B, C are not in a straight line the angles of the triangle ABC are determinate functions of the three sides, it follows that all angles are unaltered by displacements of a rigid body. If three lines meeting at a point and fixed in the body are mutually perpendicular before displacement, they are still perpendicular after the displacement.

The equations of dynamics in the form $m\ddot{x}_i = X_i$ are true with respect to dynamical (that is, non-rotating and unaccelerated) axes. Let any particle Q of a body have coordinates x_i with reference to dynamical axes at O. Then let the body be displaced in any way, and let the particle have the new coordinates x_i'. We require the relation between x_i' and x_i. Consider a particle of the body, P say, whose coordinates before and after the displacement are a_i and a_i'. Put

$$x_i = a_i + y_i; \quad x_i' = a_i' + y_i' \qquad (1).$$

Then y_i and y_i' are the coordinates of Q with respect to axes at P parallel to the dynamical axes before and after the displacement. Also if we imagine the original axes at P to be specified by the particles on them, these particles in the new position still specify a set of rectangular axes, with respect to which the coordinates of Q are still y_i. If the cosine of the angle between the y_k axis in its new position and the y_i axis in its old one is a_{ik}, we have therefore

$$y_i' = a_{ik} y_k \tag{2},$$

whence
$$x_i' = a_i' + a_{ik}(x_k - a_k) \tag{3}.$$

The displacement of Q is

$$a_i' - a_i + a_{ik} y_k - y_i \tag{4}.$$

We can prove that there is a straight line of particles in the body such that the displacement of Q is the same as that of P. For such a particle we should have

$$a_{ik} y_k - y_i = (a_{ik} - \delta_{ik}) y_k = 0 \tag{5},$$

and these three homogeneous equations in y_k are consistent provided

$$\| a_{ik} - \delta_{ik} \| = 0 \tag{6},$$

that is,
$$\begin{vmatrix} a_{11} - 1 & a_{12} & a_{13} \\ a_{21} & a_{22} - 1 & a_{23} \\ a_{31} & a_{32} & a_{33} - 1 \end{vmatrix} = 0 \tag{7}.$$

Now the determinant $\| a_{ik} \|$ is unity, and each element in it is equal to its first minor. If we expand (7) we get

$$\begin{aligned} \| a_{ik} \| - \{(a_{11}a_{22} - a_{12}a_{21}) + (a_{22}a_{33} - a_{23}a_{32}) \\ + (a_{33}a_{11} - a_{31}a_{13})\} + (a_{11} + a_{22} + a_{33}) - 1 = 0 \end{aligned} \tag{8},$$

since the terms all cancel. Hence (5) have an infinite number of solutions, all proportional, and the points therefore lie on a straight line.

If we take any two planes through this line, the angle between them is the same after displacement as before, and therefore all planes are rotated through the same angle. Thus any displacement of a rigid body is equivalent to a displacement of a particle of it combined with a rotation about a line through that particle. If the angle of rotation is θ, we have, by comparing (5) with (52) of Chapter II,

$$\begin{aligned} b_{ik} &= a_{ik} - \delta_{ik} \\ &= (l_i l_k - \delta_{ik})(1 - \cos\theta) - \epsilon_{ikm} l_m \sin\theta \end{aligned} \tag{9},$$

where l_i are the direction cosines of the axis of rotation. Thus the a_{ik} are determined in terms of this axis and the angle of rotation. Conversely, if we take the symmetrical and antisymmetrical parts,

$$(l_i l_k - \delta_{ik})(1 - \cos\theta) = \tfrac{1}{2}(b_{ik} + b_{ki}) \qquad (10),$$

$$\tfrac{1}{2}(\epsilon_{kim} - \epsilon_{ikm})\, l_m \sin\theta = \tfrac{1}{2}(b_{ik} - b_{ki}) \qquad (11).$$

But for a given m, with ikm in cyclic order,

$$\epsilon_{kim} - \epsilon_{ikm} = -2 \qquad (12).$$

Thus
$$l_m \sin\theta = -\tfrac{1}{2}(b_{ik} - b_{ki}) \qquad (13),$$

and

$$\sin^2\theta = \tfrac{1}{4}\{(b_{12} - b_{21})^2 + (b_{23} - b_{32})^2 + (b_{31} - b_{13})^2\} \quad (14).$$

Now suppose that the displacement is small. Then a_{ik} have nearly their values for zero displacement, that is, δ_{ik}. The direction cosines of the y_k and y_l axes in their new positions are a_{ik} and a_{il}, and thus, if $k \neq l$,

$$a_{ik} a_{il} = 0 \qquad (15),$$

and, if $k = l$,
$$a_{ik} a_{il} = 1 \qquad (16).$$

But in (15) for $i = k$, a_{ik} is nearly 1 and a_{il} small; for $i = l$, a_{il} is nearly 1 and a_{ik} small; for i not equal to k or l, both a_{ik} and a_{il} are small. Hence to the first order

$$a_{lk} + a_{kl} = 0 \qquad (17).$$

If $k = l$, then for $i \neq k$ or l both terms of (16) are small of the second order, and therefore for $i = k$ or l, $a_{ik} = 1 +$ a second order quantity. Thus to the first order

$$a_{ik} = \delta_{ik} - b_{ik} \qquad (18),$$

where b_{ik} is an antisymmetrical tensor.

The displacement of Q is $\delta a_i + a_{ik} y_k - y_i$, where δa_i is the displacement of P, and is a first order small quantity.

Now let X_i be the total force acting on the particle at Q. In the displacement the work it does is the scalar

$$\begin{aligned}\delta W &= X_i\,(\delta a_i + a_{ik}y_k - y_i)\\ &= X_i\,(\delta a_i - b_{ik}y_k)\\ &= X_i\delta a_i - \tfrac{1}{2}b_{ik}\,(X_i y_k + X_k y_i) - \tfrac{1}{2}b_{ik}\,(X_i y_k - X_k y_i)\end{aligned} \qquad (19).$$

When i and k are interchanged, b_{ik} and $X_i y_k - X_k y_i$ are reversed in sign, while $X_i y_k + X_k y_i$ is unaltered. Hence

$$\delta W = X_i\delta a_i + \tfrac{1}{2}\,(y_i X_k - y_k X_i)\,b_{ik} \qquad (20).$$

Now write

$$y_i X_k - y_k X_i = L_{ik} \qquad (21),$$

so that L_{ik} is an antisymmetrical tensor. The second term in (20) is the sum of nine terms, of which three are always zero and the others equal in pairs. If we replace b_{ik} and L_{ik} by the associated vectors, we have

$$\delta W = X_i\delta a_i + L_m b_m \qquad (22),$$

where the δa_i and b_m are all independent of one another and the same for all particles of the body. Hence if we add for all particles of the body

$$\delta W = (\Sigma X_i)\,\delta a_i + (\Sigma L_m)\,b_m \qquad (23),$$

and the total work done in any given small displacement is determinate if we sum up the forces acting by the six expressions ΣX_i, ΣL_m. Further, the contributions to ΣX_i, ΣL_m from the internal reactions are zero. This follows at once if these reactions consist of equal and opposite forces between pairs of particles along the line joining them, and also has the justification that it leads to correct results. Then we may restrict ΣX_i, ΣL_m to the contributions from the external forces. This is *d'Alembert's Principle*.

In the limiting case of continuous motion, we may consider the displacements that take place in a short time interval δt, and put in the limit

$$\delta a_i = u_i \delta t;\ \ b_{ik} = \omega_{ik} \delta t \qquad (24),$$

and call u_i the velocity of P and ω_{ik} the angular velocity of the body. Then the velocity of Q is

$$\dot{x}_i = u_i - \omega_{ik} y_k \qquad (25).$$

Now consider the centroid G, with coordinates $\bar{x}_i$, defined by

$$(\Sigma m)\, \bar{x}_i = \Sigma m x_i \qquad (26).$$

It is usually assumed without proof that G is fixed in the body, though this is not obvious. But suppose that the particle at G with coordinates $\bar{x}_i$ in the original position of the body goes to G' when the body is displaced, its new coordinates are $\bar{x}_i'$, given by

$$\bar{x}_i' = a_i' + a_{ik}\,(\bar{x}_k - a_k) \qquad (27),$$

and the coordinates of the new centroid G'' are $\bar{x}_i''$, given by

$$(\Sigma m)\, \bar{x}_i'' = \Sigma m x_i' \qquad (28).$$

We have to show that G' and G'' coincide. We have

$$(\Sigma m)\, \bar{x}_i'' = \Sigma m\, \{a_i' + a_{ik}\,(x_k - a_k)\} \qquad (29),$$

and therefore

$$\begin{aligned}(\Sigma m)\,(\bar{x}_i'' - \bar{x}_i') &= \Sigma m\, \{a_i' + a_{ik}\,(x_k - a_k)\} \\ &\qquad - (\Sigma m)\, \{a_i' + a_{ik}\,(\bar{x}_k - a_k)\} \\ &= \Sigma m a_{ik} x_k - \Sigma m a_{ik} \bar{x}_k \\ &= 0 \qquad (30).\end{aligned}$$

Thus the particle originally at G is displaced to the new position of the centroid, and therefore the centroid is fixed in the body.

Now return to the equations of motion of a particle of the body

$$m\ddot{x}_i = X_i \qquad (31).$$

By addition we form from these the equations

$$\Sigma m\ddot{x}_i = \Sigma X_i \qquad (32),$$

and by cross-multiplication followed by addition for all the particles of the body,

$$\Sigma m\,(x_i\ddot{x}_k - x_k\ddot{x}_i) = \Sigma\,(x_i X_k - x_k X_i) \qquad (33).$$

On the right sides of (32) and (33) the contributions from the internal reactions are zero, by d'Alembert's principle. Also

$$\Sigma m\ddot{x}_i = \frac{d^2}{dt^2}\Sigma m x_i = \frac{d^2}{dt^2}(\Sigma m)\,\bar{x}_i = M\ddot{\bar{x}}_i \qquad (34),$$

where

$$M = \Sigma m \qquad (35),$$

the total mass of the body. Also if

$$h_{ik} = \Sigma m\,(x_i\dot{x}_k - x_k\dot{x}_i) \qquad (36),$$

we can reduce (32) and (33) to

$$M\ddot{\bar{x}}_i = \Sigma X_i \qquad (37),$$

$$\frac{d}{dt}h_{ik} = \Sigma L_{ik} \qquad (38).$$

These are the fundamental equations of rigid dynamics. The three independent h_{ik} are expressible in terms of the three independent ω_{ik} by (36), and we have therefore six differential equations for the $\bar{x}_i$ and ω_{ik}, which determine them in terms of the initial conditions if the external forces are given. The motion of the body is therefore determinate.

The *principle of virtual work* follows immediately. For if a body is initially at rest, $\dot{\bar{x}}_i$ and h_{ik} are zero, and the condition that they may remain zero is that ΣX_i and ΣL_{ik} shall vanish. But this implies, by (23), that in any small displacement of the body the work done by the external forces is a small quantity of the second order in the displacements. Conversely, if there are six independent possible small displacements such that the external forces do no work, to the first order, in any of them, the coefficients of the independent δa_i and b_m of (23) must all vanish. But these coefficients are the ΣX_i and ΣL_m, and therefore the $\dot{x}_i$ and h_{ik} do not vary with the time. The vanishing of the work done by the external forces in six independent small displacements is therefore a necessary and sufficient condition for equilibrium.

The equations (38) may be put in another form. If we consider any moving origin O', not necessarily fixed in the body, with coordinates a_i, we can write for the coordinates of a particle with respect to O',

$$y_i = x_i - a_i \qquad (39).$$

If any vector associated with the particle, such as its velocity, momentum, or acceleration, or the force acting on it, has components u_i, we may form the antisymmetrical tensor $y_i u_k - y_k u_i$ and call it the *moment* of the vector about O'. From (32) and (33) we can form the equations

$$\Sigma m\,(x_i \ddot{x}_k - x_k \ddot{x}_i) - (a_i \Sigma m \ddot{x}_k - a_k \Sigma m \ddot{x}_i)$$
$$= \Sigma\,(x_i X_k - x_k X_i) - (a_i \Sigma X_k - a_k \Sigma X_i) \qquad (40),$$

where only the external forces make any contribution to the right side. But by (39) this is equivalent to

$$\Sigma m\,(y_i \ddot{x}_k - y_k \ddot{x}_i) = \Sigma\,(y_i X_k - y_k X_i) \qquad (41).$$

Therefore the sum of the moments of the mass-acceleration products about *any* origin is equal to the sum of the moments of the external forces about that origin.

If we denote the moment of momentum, or angular momentum, about O' by h_{ik}', we have

$$h_{ik}' = \Sigma m\,(y_i \dot{x}_k - y_k \dot{x}_i) \qquad (42),$$

and

$$\frac{d}{dt} h_{ik}' = \Sigma m\,(y_i \ddot{x}_k - y_k \ddot{x}_i) + \Sigma m\,(\dot{y}_i \dot{x}_k - \dot{y}_k \dot{x}_i) \qquad (43).$$

The second term may be written

$$\Sigma m \{(\dot{x}_i - \dot{a}_i)\,\dot{x}_k - (\dot{x}_k - \dot{a}_k)\,\dot{x}_i\} = -\,\Sigma m\,(\dot{a}_i\dot{x}_k - \dot{a}_k\dot{x}_i)$$

$$= -\,M\,(\dot{a}_i\dot{\bar{x}}_k - \dot{a}_k\dot{\bar{x}}_i) \quad (44).$$

Thus (41) is equivalent to

$$\frac{d}{dt}h_{ik}' + M\,(\dot{a}_i\dot{\bar{x}}_k - \dot{a}_k\dot{\bar{x}}_i) = \Sigma\,(y_i X_k - y_k X_i) \quad (45).$$

In many important cases the second tensor on the left vanishes identically. This is clearly true if O' is fixed, when the time derivatives of the a_i are zero; when the centroid is fixed; and, if both are moving, if the vectors $\dot{a}_i$ and $\dot{\bar{x}}_i$ are parallel, that is, if the velocity of the moving origin O' is parallel to that of the centroid. The most important case is where the moving origin is identical with the centroid, when the last condition holds automatically. These terms also disappear if the moving origin is an instantaneous centre of rotation always at the same distance from the centroid. They vanish for a sphere or circular cylinder rolling down an inclined plane, but not for a rolling elliptic cylinder.

If the moving origin is the centroid,

$$h_{ik}' = \Sigma m\,\{y_i\,(\dot{\bar{x}}_k + \dot{y}_k) - y_k\,(\dot{\bar{x}}_i + \dot{y}_i)\}$$

$$= \Sigma m\,(y_i\dot{y}_k - y_k\dot{y}_i) \quad (46),$$

since $\Sigma m y_i = \Sigma m y_k = 0,$

by the definition of the centroid. Thus the angular momenta about the centroid are expressed completely in terms of positions and motions relative to the centroid, and the formulae for them have the same form as those for the total angular momenta with reference to a fixed origin.

By (25) $\dot{y}_i = -\,\omega_{im}y_m \quad (47),$

m being here a dummy suffix. Substituting in (46) we have

$$h_{ik}' = -\,\Sigma m\,(\omega_{km}y_i y_m - \omega_{im}y_k y_m)$$

$$= B_{km}\omega_{im} - B_{im}\omega_{km} \quad (48),$$

where B_{im} is the symmetrical tensor defined by

$$B_{im} = \Sigma m y_i y_m \qquad (49),$$

and depends only on the masses and positions of the particles of the body. In the three-dimensional case, with i and k unequal, m must be equal in turn to i, k, and the other value. Then, for instance,

$$\begin{aligned} h_{12}' &= (B_{21}\omega_{11} + B_{22}\omega_{12} + B_{23}\omega_{13}) \\ &\qquad - (B_{11}\omega_{21} + B_{12}\omega_{22} + B_{13}\omega_{23}) \\ &= (B_{22} + B_{11})\,\omega_{12} - B_{13}\omega_{23} - B_{23}\omega_{31} \qquad (50). \end{aligned}$$

If we replace the antisymmetrical tensors by the associated vectors this takes the form

$$\begin{aligned} h_3' &= (B_{11} + B_{22})\,\omega_3 - B_{13}\omega_1 - B_{23}\omega_2 \\ &= (B_{11} + B_{22} + B_{33})\,\omega_3 - (B_{13}\omega_1 + B_{23}\omega_2 + B_{33}\omega_3) \quad (51), \end{aligned}$$

and in general

$$h_m' = A_{im}\omega_i \qquad (52),$$

where A_{im} is the symmetrical tensor given by

$$\begin{aligned} A_{im} &= B_{kk}\delta_{im} - B_{im} \\ &= (\Sigma m y_k^2)\,\delta_{im} - \Sigma m y_i y_m \qquad (53). \end{aligned}$$

It evidently corresponds to the system of moments and products of inertia given in ordinary treatises on dynamics; A_{11}, A_{22}, A_{33} are the ordinary moments of inertia A, B, C, but A_{23}, A_{31}, A_{12} are equal and opposite to the ordinary products of inertia F, G, H.

It should be noticed that this reduction is characteristic of three dimensions; in a higher number of dimensions there is no analogous simplification of the form (48).

The equations of motion of a rigid body then take the form

$$M\ddot{x}_i = \Sigma X_i \qquad (54),$$

$$\frac{d}{dt} h_{ik}' = \Sigma L_{ik}'; \text{ or } \frac{d}{dt} h_m' = \Sigma L_m' \qquad (55),$$

where L_{ik}' is the moment of the external forces about the centroid.

It may happen that one point of the body is fixed. In that case we may take the origin at that point. Then in addition to the known external forces there is a reaction at the origin which can if required be found from (37); but the reaction has no moment about the origin and the motion is given by (38). But in this case

$$\dot{x}_i = -\omega_{im} x_m \qquad (56),$$

and we find, by a process analogous to the last one,

$$h_{ik} = D_{km}\omega_{im} - D_{im}\omega_{km} \qquad (57),$$

where

$$D_{im} = \Sigma m x_i x_m \qquad (58).$$

In terms of the associated vector,

$$h_m = C_{im}\omega_i \qquad (59),$$

where

$$C_{im} = (\Sigma m x_k^2)\,\delta_{im} - \Sigma m x_i x_m \qquad (60).$$

The C_{im} correspond to the moments and products of inertia about the origin. They can be expressed in terms of those about the centroid; for

$$\begin{aligned} C_{im} &= \{\Sigma m\,(\bar{x}_k + y_k)^2\}\,\delta_{im} - \Sigma m\,(\bar{x}_i + y_i)\,(\bar{x}_m + y_m) \\ &= (\Sigma m \bar{x}_k^2)\,\delta_{im} - \Sigma m \bar{x}_i \bar{x}_m \\ &\quad + A_{im} \qquad (61), \end{aligned}$$

the terms linear in the y's vanishing by the definition of the centroid.

The relevant equations of motion then take the form

$$\frac{d}{dt} h_m = \Sigma L_m \qquad (62),$$

where the L_m are the moments of the external forces about the origin, the reaction at the origin making no contribution.

The kinetic energy of the body is given by

$$\begin{aligned} 2T &= \Sigma m \dot{x}_i^2 \\ &= \Sigma m\,(\dot{\bar{x}}_i - \omega_{im} y_m)^2 \\ &= \Sigma m \dot{\bar{x}}_i^2 + \Sigma m y_k y_m \omega_{ik} \omega_{im} \\ &= M \dot{\bar{x}}_i^2 + B_{km} \omega_{ik} \omega_{im} \qquad (63). \end{aligned}$$

In three dimensions $\omega_{ik}\omega_{im}$ is zero unless i is different from both k and m. If $i = 1$, $k = 2$, $m = 3$, or if $i = 1$, $k = 3$, $m = 2$, $\omega_{12}\omega_{13} = -\omega_2\omega_3$. If $i = 1$, $k = m = 2$, $\omega_{12}^2 = \omega_3^2$, and has coefficient B_{22}. Thus ω_3^2 enters with a coefficient $B_{22} + B_{11}$, or A_{33}. But $\omega_2\omega_3$ has a coefficient $-2B_{23}$ or $A_{23} + A_{32}$. Thus in all

$$2T = M\dot{\bar{x}}^2 + A_{km}\omega_k\omega_m \qquad (64).$$

When a point of the body is kept fixed,

$$2T = C_{km}\omega_k\omega_m \qquad (65).$$

We notice that the linear and angular momenta take the forms

$$M\ddot{\bar{x}}_i = \frac{\partial T}{\partial \dot{x}_i}; \quad h_i' = \frac{\partial T}{\partial \omega_i} \ (\dot{\bar{x}}_k \text{ constant}),$$

$$h_i = \frac{\partial T}{\partial \omega_i} \qquad (66)$$

for origin at a point of the body held fixed.

In one respect the forms (55) and (62) are inconvenient. They involve the tensors A_{im} and C_{im}, which depend on the x_i and y_i and therefore in general change as the body rotates. It is more convenient to use such axes that the relevant tensor in the particular problem is constant. To achieve this the axes must rotate, and then are no longer dynamical axes. Suppose then that we have a set of axes x_j' rotating in any way, and that their direction cosines with respect to the dynamical axes x_i are a_{ij}. All the usual tensor relations hold for transformations from the x_j'

system to the x_i system and *vice versa*; but the a_{ij}, while specified for each instant, are now functions of the time. If u_i is any vector, we have

$$u_j' = a_{ij}u_i; \quad u_i = a_{ij}u_j' \tag{67}$$

Then $\dfrac{du_i}{dt}$ is another vector; and

$$\frac{du_i}{dt} = a_{ij}\frac{du_j'}{dt} + u_j'\frac{da_{ij}}{dt} \tag{68}$$

The component of this vector in the x_l' direction is

$$a_{il}a_{ij}\frac{du_j'}{dt} + a_{il}u_j'\frac{da_{ij}}{dt} \tag{69}$$

The first term of this is

$$\delta_{lj}\frac{du_j'}{dt} = \frac{du_l'}{dt} \tag{70}$$

Also $\dfrac{da_{ij}}{dt}$ is the x_i velocity of a point at unit distance along the x_j' axis, that is, a point with coordinates a_{ij} $(i = 1, 2, 3)$ with reference to the dynamical axes. This velocity is $-\theta_{ik}a_{kj}$, where θ_{ik} is the antisymmetrical tensor expressing the rotation of the rigid frame consisting of the moving axes. Then

$$\begin{aligned} a_{il}u_j'\frac{da_{ij}}{dt} &= -a_{il}u_j'\theta_{ik}a_{kj} \\ &= -\theta_{lj}'u_j' \end{aligned} \tag{71}$$

and the required component is

$$\frac{du_l'}{dt} - \theta_{lj}'u_j' \tag{72}$$

If we use instead of θ_{jl}' the associated vector, the three components become

$$(\dot{u}_1' - u_2'\theta_3' + u_3'\theta_2',\ \dot{u}_2' - u_3'\theta_1' + u_1'\theta_3',\ \dot{u}_3' - u_1'\theta_2' + u_2'\theta_1') \tag{73}$$

If for instance the components of a displacement along the x_j' axes are u_j', the formula (73) gives the components of the velocity along these axes; if u_j' are the components of velocity, (73) gives the components of acceleration; if u_j' are the components of angular momentum, (73) gives the components of their rates of change with reference to dynamical axes, and these are equal to the components of the moments of the forces acting.

EXAMPLE

Prove that

$$h_p' = -\Sigma m y_i y_k \epsilon_{imp} \epsilon_{mks} \omega_s$$
$$= A_{ip} \omega_i .$$

CHAPTER V

EQUIVALENCE OF SYSTEMS OF FORCES

An external force X_i acting at a point x_i of a rigid body produces dynamical effects summed up in the vector X_i and the antisymmetrical tensor $x_i X_k - x_k X_i$. A force has a line of action; that is, if we take its resultant R given by

$$R^2 = X_i^2 \quad (1)$$

we can define a direction l_i by

$$X_i = Rl_i \quad (2).$$

By convention R is always taken positive. If a force X_i acts at the point $x_i + rl_i$, where r is arbitrary, we have

$$(x_i + rl_i) X_k - (x_k + rl_k) X_i = x_i X_k - x_k X_i \quad (3).$$

Thus the dynamical effects are the same if the force X_i acts at *any* point of a line through x_i with direction cosines proportional to X_i. This line is called the *line of action* of the force, and the force can be said to act *along* it.

If a force has magnitude R and acts at x_i in the direction l_i, we have

$$X_i = Rl_i \quad (4),$$

$$L_{ik} = R(x_i l_k - x_k l_i) \quad (5),$$

or

$$L_m = Rl_m' \quad (6),$$

in the notation of the coordinates of a line. Thus the X_i and L_m are the products of the resultant into the six coordinates of the line of action.

The moment of the force X_i about a point a_i is

$$(x_i - a_i) X_k - (x_k - a_k) X_i = R(l_{ik}' - a_i l_k + a_k l_i) \quad (7).$$

If we take the associated vector, its component in the direction m_m is

$$R\,(m_m l_m' - \epsilon_{ikm} m_m a_i l_k) = R\,(m_i l_i' + l_i m_i') \quad (8),$$

where m_i' are the other coordinates of the line through a_i in the direction m_i. We notice that

$$m_i l_i' + l_i m_i' = d \sin\theta \quad (9),$$

where θ is the angle between the two lines and d the length of their common perpendicular. We may call $Rd\sin\theta$ the moment of the force about the *line* (m_i, m_i').

By d'Alembert's principle, the motion of a body is unaltered if to the forces acting on it we add two equal and opposite forces acting at the same point, or, by (3), along the same line of action.

Now consider a pair of equal and opposite forces X_i and $-X_i$ acting at points a_i and b_i. They clearly make no contribution to ΣX_i. Their contribution to ΣL_{ik} is

$$(a_i X_k - a_k X_i) - (b_i X_k - b_k X_i)$$
$$= (a_i - b_i)\,X_k - (a_k - b_k)\,X_i \quad (10).$$

Since a_i and b_i are equally affected by any motion of the origin, the contribution of such a pair of forces applied to definite particles to both ΣX_i and ΣL_{ik} is independent of the position of the origin. Such a pair is called a *couple*, and its contribution to ΣL_{ik} is called the *moment of the couple*.

If the vectors $a_i - b_i$ and X_i are both perpendicular to a line with direction cosines n_i, the components of ΣL_m are the components of a vector along this line; while the forces act in the same plane perpendicular to the line. This plane is called the *plane of the couple*, and the line an *axis* of the couple. Evidently equal couples in parallel planes are equivalent. The magnitude of this vector is Rd, where $R^2 = X_i^2$ and d is the perpendicular distance of a_i from a line through b_i parallel to X_i; and its components can accordingly be written Rdn_m.

Any system of forces is equivalent to a force at an arbitrary point a_i together with a couple. For with each X_i acting at x_i we can associate a pair of forces $\pm X_i$ at a_i. Then the system is equivalent to ΣX_i at a_i together with a set of couples whose total moment is

$$M_{ik} = \Sigma \{(x_i - a_i) X_k - (x_k - a_k) X_i\} \qquad (11).$$

But this is equivalent to a single couple; for we have only to make n_m proportional to the M_m and Rd equal to their resultant.

If L_{ik} are the moments about the origin,

$$L_{ik} = \Sigma (x_i X_k - x_k X_i) \qquad (12),$$

and if for brevity we replace ΣX_i by simply X_i,

$$M_{ik} = L_{ik} - (a_i X_k - a_k X_i) \qquad (13),$$

or

$$M_m = L_m - \epsilon_{ikm} a_i X_k \qquad (14).$$

Evidently X_i^2 is a scalar and independent of a_i. Also

$$X_m M_m = X_m L_m - \epsilon_{ikm} a_i X_k X_m \qquad (15).$$

The first term is a scalar and independent of a_i. The second is identically zero. Hence X_i^2 and $X_i M_i$ are scalar invariants.

The system is equivalent to a force X_i and a couple M_i at a_i. These vectors are parallel if

$$L_m - \epsilon_{ikm} a_i X_k = pX_m \qquad (16),$$

where p is a scalar length called the *pitch*. These give three linear relations between the three a_i and the pitch, and we therefore expect a single infinity of solutions. But if we take the scalar product of (16) by X_m we have

$$X_m L_m = pX_m^2 \qquad (17),$$

so that p is the same for all admissible values of a_i. With this value of p, (16) represents three planes with a line in common. If we change a_i to $a_i + \sigma X_i$, where σ is a scalar, the left side of (16) is increased by

$$-\epsilon_{ikm}\sigma X_i X_k = 0 \tag{18},$$

and therefore if a_i is one point on the line, all points on the line through a_i parallel to X_i satisfy the conditions. This line is the *central axis* of the system; the system is equivalent to a force R along the central axis and a couple G about it. This expresses the system as a *wrench*. If we take the central axis as one of the coordinate axes, we have, since X_i^2 and $X_i M_i$ are scalars,

$$R^2 = X_i^2 \tag{19},$$

$$GR = X_i M_i = pX_i^2 \tag{20},$$

and therefore G, R, and p are determined.

The system can also be reduced to a couple parallel to a preassigned plane together with a force. For if S is the couple, and n_i are the direction cosines of the normal to the plane, and if the force acts through a_i, we have, for the moments about a_i,

$$M_m = L_m - \epsilon_{ikm} a_i X_k \tag{21},$$

and also

$$M_m = Sn_m \tag{22}.$$

We have three equations to determine the a_i and S. Again there are a single infinity of solutions. But if we take the scalar product by X_m, we get

$$\begin{aligned} Sn_m X_m &= L_m X_m - \epsilon_{ikm} a_i X_k X_m \\ &= GR \end{aligned} \tag{23},$$

so that S is determined provided $n_m X_m$ is not zero, that is, provided the resultant force is not parallel to the plane. Then the equations

$$L_m - \epsilon_{ikm} a_i X_k = Sn_m \tag{24}$$

determine a line parallel to X_i, which is the line of action of the force.

If l_i are the direction cosines of a line through a_i, the moment of the system about this line is

$$l_m M_m = l_m L_m - \epsilon_{ikm} a_i X_k l_m \quad (25)$$

$$= l_m L_m + l_k' X_k \quad (26),$$

where l_k' are the other coordinates of the line. If this moment vanishes the line is called a *null line* of the system. If b_i is another point on it,

$$L_m (b_m - a_m) = \epsilon_{ikm} a_i X_k (b_m - a_m) \quad (27),$$

which shows that b_i lies in a definite plane through a_i. All null lines through a point therefore lie in one plane. This plane is called the *null plane* of the point.

All null lines in a plane pass through a point. For the system can in general be reduced to a couple in the plane and a force whose line of action intersects the plane in one point. Then the system has no moment about any line in the plane through this point, which is the *null point* of the plane.

Any system is equivalent to two forces, one of which can be made to act along a given line. For let the lines of action pass through a_i and b_i, and have direction cosines l_i and m_i, and let the magnitudes of the forces be S and T. Then we have six equations,

$$X_i = Sl_i + Tm_i \quad (28),$$

$$L_m = \epsilon_{ikm} (a_i Sl_k + b_i Tm_k)$$

$$= Sl_m' + Tm_m' \quad (29).$$

The coordinates of the first line being given, these are six linear equations to determine the six coordinates of the second line and S and T. But we have also

$$m_i^2 = 1; \; m_i m_i' = 0 \quad (30),$$

so that we have in general just enough equations. Two such lines are *conjugate lines*. Clearly any line intersecting two conjugate lines is a null line.

It can be shown easily that

$$S\,(X_i l_i' + L_i l_i) = L_i X_i \qquad (31),$$

$$T^2 = X_i^2 - 2Sl_i X_i + S^2 \qquad (32),$$

whence (28) and (29) determine the coordinates of the second line explicitly.

EXAMPLES

1. A system of forces is reduced to a force at P together with a couple; P is chosen so that the couple is parallel to a given plane. Show that the locus of P is a straight line parallel to the central axis.

2. Show that

$$a_m = \frac{\epsilon_{ikm} X_i L_k}{X_i^2}$$

is a point on the central axis.

3. Two systems of forces are given by $(X_i,\ L_i), (Y_i,\ K_i)$. Show that $X_i K_i + Y_i L_i$ is invariant.

CHAPTER VI

CONTINUOUS SYSTEMS

In problems involving volume and surface integrals we find it convenient to denote elements of volume and surface by $d\tau$ and dS respectively; both are always taken positive. They are of course scalars. The direction cosines of the normal to an element of surface are usually denoted by l_i; in most cases the normal is drawn *outwards* from the region under consideration.

Green's Lemma takes the form

$$\iiint \frac{\partial u_i}{\partial x_i}\, d\tau = \iint l_i u_i\, dS \tag{1},$$

and we have the corollary, if

$$u_i = \partial V/\partial x_i \tag{2},$$

where V is a scalar,

$$\iiint \nabla^2 V\, d\tau = \iint \frac{\partial V}{\partial n}\, dS \tag{3},$$

where

$$\nabla^2 = \frac{\partial^2}{\partial x_i^{\,2}} = \frac{\partial^2}{\partial x_1^{\,2}} + \frac{\partial^2}{\partial x_2^{\,2}} + \frac{\partial^2}{\partial x_3^{\,2}} \tag{4},$$

and $\partial/\partial n$ denotes differentiation along the outward normal.

Stokes's Theorem takes the form

$$\begin{aligned}\int_C u_i\, dx_i &= \iint \left\{ l_1 \left(\frac{\partial u_3}{\partial x_2} - \frac{\partial u_2}{\partial x_3} \right) + l_2 \left(\frac{\partial u_1}{\partial x_3} - \frac{\partial u_3}{\partial x_1} \right) \right. \\ &\qquad \left. + l_3 \left(\frac{\partial u_2}{\partial x_1} - \frac{\partial u_1}{\partial x_2} \right) \right\} dS \\ &= \iint l_i \epsilon_{ikm}\, \partial u_m/\partial x_k\, dS \end{aligned} \tag{5}.$$

The integral on the left is round a closed contour C. On the right l_i is the direction cosine of the normal to any element dS of a surface S that fills up the contour; the integral is over S; and the sense of the normal is such that if the contour is described in the positive sense about any axis x_i, l_i is taken positive when the normal is in the direction of x_i increasing.

The gravitational potential of a distribution of matter is given by

$$V = f\Sigma \frac{m}{r} \qquad (6),$$

where m is an element of mass, r is its distance from the point where V is to be found, and f is a constant equal to $6{\cdot}66 \times 10^{-8}$ when m and r are measured in grams and centimetres and the time in seconds. When the mass is distributed continuously over a surface or through a volume, m must be replaced by σdS or $\rho d\tau$ respectively, where σ and ρ are called the surface and volume densities. The work in displacing a mass m' through a small distance is $m'dV$.

The electrostatic potential of a system of point charges is given by

$$V = f\Sigma \frac{e}{r} \qquad (7),$$

where e is a typical element of charge. If e is in electrostatic units of charge and r in centimetres, f is $+1$. The work in a small displacement of a charge e' in the field is $-e'dV$.

The usual relations follow, that in free space in both cases

$$\nabla^2 V = 0 \qquad (8),$$

and in space occupied by matter of finite density

$$\nabla^2 V = -4\pi f\rho \qquad (9).$$

Also we have Gauss's Theorem

$$\iint \frac{\partial V}{\partial n}\, dS = -4\pi f\Sigma' m \text{ or } -4\pi f\Sigma' e \qquad (10),$$

where the summation is over all the masses or charges within the closed surface S. In crossing a surface where there is a finite surface density σ, $\partial V/\partial n$ has a finite discontinuity $-4\pi f\sigma$.

All these relations are scalar in form. For a gravitational field the force on a small particle of mass m' is

$$m'X_i = m'\partial V/\partial x_i \qquad (11),$$

while for an electrostatic field the force on a small charge e' is

$$e'X_i = -e'\partial V/\partial x_i \qquad (12).$$

The vectors X_i are called the *intensities* of gravitational and electric force respectively. The analogy in form between (6) and (7) is constructed for mathematical convenience; the difference in sign in (11) and (12) embodies the physical difference that whereas two positive masses attract, two positive charges repel.

In a gravitating system we may construct a work-function

$$W = \Sigma_p \Sigma_q f \frac{m_p m_q}{r_{pq}} \qquad (13),$$

where the summation is over all pairs of particles m_p, m_q, and r_{pq} is the distance between m_p and m_q. The case where $p = q$ is excluded. Then the force on the pth particle is $\partial W/(\partial x_i)_p$. The potential at m_p due to the other particles is $V_p = \Sigma' f m_q/r_{pq}$, the accent indicating that the case where $q = p$ is excluded from the summation; and the function $\Sigma m_p V_p = 2W$, since each pair of particles is counted twice in this double summation. The function $\frac{1}{2}\Sigma m_p V_p$ therefore plays the part of a work-function. Similarly in electrostatics the function $\frac{1}{2}\Sigma e_p V_p$ plays the part of the work-function with its sign reversed, that is, of a potential energy. These results may be generalized to the case of continuous distributions; thus we can replace these functions by

$$W = \tfrac{1}{2}\iiint \rho V\, d\tau + \tfrac{1}{2}\iint \sigma V\, dS \qquad (14),$$

where the first integral is through all regions where ρ is finite and the second over all surfaces where σ is finite; the first integral may therefore be taken through all space, excluding the surfaces where there is a surface density. It can be shown that with proper precautions about the definition of V the restriction $p \neq q$ gives no trouble provided V is everywhere finite.

Now if we consider the integral through all space except thin laminae surrounding the surfaces where σ is finite,

$$\begin{aligned}\tfrac{1}{2}\iiint \rho V \, d\tau &= -\frac{1}{8\pi f}\iiint V \frac{\partial^2 V}{\partial x_i^2}\, d\tau \\ &= -\frac{1}{8\pi f}\iiint \left\{\frac{\partial}{\partial x_i}\left(V\frac{\partial V}{\partial x_i}\right) - \left(\frac{\partial V}{\partial x_i}\right)^2\right\} d\tau \\ &= -\frac{1}{8\pi f}\iint V \frac{\partial V}{\partial n}\, dS + \frac{1}{8\pi f}\iiint\left(\frac{\partial V}{\partial x_i}\right)^2 d\tau \quad (15),\end{aligned}$$

where the dn in the first integral is out from the region of integration and therefore towards the surface where σ is finite. On the two sides of such a surface the values of V differ by an indefinitely small amount, and for the two sides together $\frac{\partial V}{\partial n}\, dS = 4\pi f\sigma dS$, by Gauss's Theorem. Hence the first integral in (15) is equal to $-\tfrac{1}{2}\iint \sigma V \, dS$ taken over the surface and cancels the second term in (14). Hence

$$W = \frac{1}{8\pi f}\iiint\left(\frac{\partial V}{\partial x_i}\right)^2 d\tau \qquad (16)$$

through all space. In consequence of this form we may say that the gravitational work-function, or the electrostatic potential energy, is $R^2/8\pi f$ per unit volume, where R is the resultant of the appropriate intensity vector.

When the properties of the medium vary from place to place, V is no longer of the form $\Sigma fe/r$, and $\nabla^2 V$ is no longer zero. But a potential still exists; if a small charge e' is moved from a point P to a point Q, the work done is still the same whatever the route, and may be denoted by $e'(V_P - V_Q)$. The treatment is suggested by the fact that two similar condensers with the plates at the same potentials, but with air between the plates of one and another material between those of the other, have charges in a ratio K depending only on the media. $\partial V/\partial n$ on the outside of the condenser being small compared with its value between the plates, we infer that the charge per unit area for the same distribution of V is related to the discontinuity in $K\partial V/\partial n$, where K depends on the material. This suggests in turn that Gauss's Theorem must be replaced by

$$\iint K \frac{\partial V}{\partial n}\, dS = -4\pi f \Sigma e \qquad (17),$$

where the summation is for all charges inside S. Then applying this to the two sides of any surface we have

$$\left[K \frac{\partial V}{\partial n}\right] = -4\pi f \sigma \qquad (18),$$

and applying it to a region with a finite volume density we have

$$-4\pi f \iiint \rho\, d\tau = \iiint \frac{\partial}{\partial x_i}\left(K \frac{\partial V}{\partial x_i}\right) d\tau \qquad (19),$$

and as this must hold for all such regions,

$$\frac{\partial}{\partial x_i}\left(K \frac{\partial V}{\partial x_i}\right) = -4\pi f \rho \qquad (20).$$

These equations are all homogeneous in V, ρ, σ; hence the potential due to any set of charges is proportional to the charges if all are altered in the same ratio. Using this principle we can show by the usual method that the energy of a distribution is

$$\tfrac{1}{2}\Sigma Ve = \tfrac{1}{2}\iiint \rho V\, d\tau + \tfrac{1}{2}\iint \sigma V\, dS \qquad (21).$$

The first integral, applied to all space except thin regions to cut off surfaces where there are surface densities, gives

$$-\frac{1}{8\pi f}\iiint V\frac{\partial}{\partial x_i}\left(K\frac{\partial V}{\partial x_i}\right)d\tau$$

$$=-\frac{1}{8\pi f}\iiint\left\{\frac{\partial}{\partial x_i}\left(KV\frac{\partial V}{\partial x_i}\right)-K\left(\frac{\partial V}{\partial x_i}\right)^2\right\}d\tau$$

$$=-\frac{1}{8\pi f}\iint KV\frac{\partial V}{\partial n}\,dS+\frac{1}{8\pi f}\iiint K\left(\frac{\partial V}{\partial x_i}\right)^2 d\tau \quad (22).$$

The surface integral cancels the integral $\frac{1}{2}\iint \sigma V\,dS$; and hence

$$W=\frac{1}{8\pi f}\iiint K\left(\frac{\partial V}{\partial x_i}\right)^2 d\tau \quad (23),$$

so that the energy can be considered equal to $KR^2/8\pi f$ per unit volume.

Magnetism may be treated similarly, starting from the assumption of volume and surface distributions of magnetic pole strength, subject to the condition that the total pole strength in any solid is zero; or we may regard the ultimate magnetic unit as the doublet, which explains the need for the restriction involved in the former method of treatment. The potential at x_i due to a doublet of strength M at the origin with its axis in the direction λ_i is

$$V=\gamma M\,\lambda_i x_i/r^3 \quad (24),$$

and if the doublet strength per unit volume in a solid is I in the direction λ_i we can introduce the intensity of magnetization at ξ_i, the vector $A_i = I\lambda_i$, and say that the potential at x_i is

$$V=\gamma\iiint A_i\frac{x_i-\xi_i}{r^3}\,d\tau \quad (25),$$

where A_i corresponds to the point ξ_i and $d\tau = d\xi_1 d\xi_2 d\xi_3$; γ is a constant. The magnetic force in free space is

$$\alpha_i=-\frac{\partial V}{\partial x_i} \quad (26).$$

We may write

$$V = \gamma \iiint A_i \frac{\partial}{\partial \xi_i}\left(\frac{1}{r}\right) d\tau$$

$$= \gamma \iiint \left\{\frac{\partial}{\partial \xi_i}\left(\frac{A_i}{r}\right) - \frac{1}{r}\frac{\partial A_i}{\partial \xi_i}\right\} d\tau$$

$$= \gamma \iint \frac{l_i A_i}{r}\, dS - \gamma \iiint \frac{\partial A_i}{\partial \xi_i}\frac{1}{r}\, d\tau \qquad (27).$$

The potential is therefore equivalent to that due to a distribution of magnetic poles $l_i A_i$ per unit area over the boundary and $-\partial A_i/\partial \xi_i$ per unit volume through the interior.

Within a solid special treatment is needed. To define V or the force at x_i, when x_i is within a solid, we must imagine a small cavity made about x_i, the intensity of magnetization everywhere remaining as before, and consider V and α_i within it; then the values of V and α_i at x_i are defined to be the limits of those in the cavity when the dimensions of the cavity become indefinitely small. This process leads to little difficulty in gravitational and electrostatic problems, but in magnetism the limit of the force is found to depend on the shape and orientation of the cavity. The force in the cavity can be written

$$X_i = -\frac{\partial V}{\partial x_i} \qquad (28),$$

where V is given by (27); in the first integral the normal is inwards towards the cavity. The contributions to X_i from the volume integral and the outer boundary are of the same form as for gravitation, and give no trouble. If the cavity is a cylinder in the direction of the intensity of magnetization, $l_i A_i$ is zero over the sides and equal to I, the resultant intensity of magnetization, on the ends. Such a surface density over the ends in the limit contributes nothing to V; if the radius of the cylinder is small compared to its length it also contributes nothing to X_i; but if the cylinder is of disc-like form it contributes $4\pi\gamma\lambda_i I$ to X_i. Hence if we take V for the complete body and define α_i by

$$\alpha_i = -\frac{\partial V}{\partial x_i} \qquad (29),$$

α_i is the value taken by X_i in a thin cylindrical cavity parallel to the intensity of magnetization. The force in a flat cylindrical cavity with its generators in this direction is

$$a_i = \alpha_i + 4\pi\gamma\lambda_i I = \alpha_i + 4\pi\gamma A_i \qquad (30).$$

Evidently α_i and a_i are both vectors; the former is called the magnetic force and the latter the magnetic induction. The theory of susceptibility and permeability may then be developed as usual. Also (27) shows that V is continuous across a boundary; but $l_i\alpha_i = \partial V/\partial n$ has a discontinuity $-4\pi\gamma l_i A_i$; whence $l_i a_i$ is continuous across a boundary.

The mutual potential energy of two doublets M and M' at x_i and x_i', oriented in directions λ_i and λ_i', is

$$W = \lambda_i' M' \frac{\partial V}{\partial x_i'} \qquad (31),$$

where V is the potential at x_i' due to the magnet at x_i; this gives

$$W = \gamma M M' \lambda_i' \frac{\partial}{\partial x_i'}\left(\lambda_k \frac{x_k' - x_k}{r^3}\right)$$

$$= \gamma M M' \left\{\frac{\lambda_i'\lambda_i}{r^3} - 3\frac{\lambda_i'\lambda_k}{r^3}\frac{x_k' - x_k}{r}\frac{x_i' - x_i}{r}\right\} \qquad (32)$$

$$= \frac{\gamma M M'}{r^3}(\cos\epsilon - 3\cos\theta\cos\theta') \qquad (33),$$

where ϵ is the angle between the axes of the magnets and θ and θ' are the angles made by the axes with the line joining the centres.

If the second magnet is turned through a small angle $\delta\psi$ about a line with direction cosines n_i,

$$\delta\lambda_i' = \epsilon_{ikm} n_k \lambda_m' \, \delta\psi \qquad (34),$$

by (49) of Chapter II. Hence, by (32),

$$\delta W = \frac{\partial W}{\partial \lambda_i'} \delta\lambda_i'$$

$$= \gamma M M' \frac{\delta\psi}{r^3} \left\{ \lambda_i - 3 \cos\theta \frac{x_i' - x_i}{r} \right\} \epsilon_{ikm} n_k \lambda_m',$$

so that the couple about a line parallel to the axis of x_k is

$$M_k = \frac{\gamma M M'}{r^3} \epsilon_{ikm} \lambda_m' \left\{ \lambda_i - 3 \cos\theta \frac{x_i' - x_i}{r} \right\} \qquad (35).$$

Hydrostatics and Classical Hydrodynamics. The internal reaction in a fluid across an element of surface dS is a pressure $p\,dS$ normal to that surface. If the density is ρ, the bodily force per unit mass X_i, and the velocity of the fluid at x_i is u_i, the acceleration of the fluid is found, by considering a small parallelepiped, to be given by

$$\rho \frac{du_i}{dt} = -\frac{\partial p}{\partial x_i} + \rho X_i \qquad (1).$$

If u_i is given in the Eulerian way as a function of the coordinates x_i and the time t, the operator d/dt, giving the rate of change of any element associated with a given particle of the fluid, is equivalent to

$$\frac{d}{dt} = \frac{\partial}{\partial t} + u_k \frac{\partial}{\partial x_k} \qquad (2).$$

If we consider the circulation Ω around any closed circuit C in the fluid and moving with it, defined by

$$\Omega = \int_C u_i \, dx_i \qquad (3),$$

we have

$$\frac{d\Omega}{dt} = \int_C \frac{du_i}{dt} dx_i + \int_C u_i \frac{d}{dt}(dx_i)$$

$$= -\int_C \frac{1}{\rho}\frac{\partial p}{\partial x_i} dx_i + \int_C X_i dx_i + \int_C u_i du_i \qquad (4).$$

The last integral is $[\frac{1}{2}u_i^2]$, which always vanishes because when we move round the contour we come back to the same point, where the velocity has its original value. Also if X_i is the gradient of a single-valued potential, as when the bodily forces are due to gravity (the commonest case), $\int X_i dx_i$ is the change of this potential round the contour and is zero. Again, if ρ is a function of p only, as in an incompressible liquid or a gas at uniform temperature, the first integral vanishes and

$$\frac{d\Omega}{dt} = 0 \qquad (5).$$

If then Ω is ever zero around a circuit it remains so permanently. This is true if the fluid is initially at rest and is set in motion by solids moving in it, and in various other cases of importance. But the vanishing of Ω for all circuits is the condition for the existence of a velocity potential ϕ such that

$$u_i = \frac{\partial \phi}{\partial x_i} \qquad (6).$$

In this case we can rewrite the equations of motion in the form

$$\frac{\partial u_i}{\partial t} + u_k \frac{\partial u_i}{\partial x_k} = -\frac{1}{\rho}\frac{\partial p}{\partial x_i} + X_i \qquad (7),$$

and multiplying by dx_i and adding we have

$$\frac{\partial}{\partial t}\frac{\partial \phi}{\partial x_i} dx_i + u_k \frac{\partial u_k}{\partial x_i} dx_i = -\frac{1}{\rho}\frac{\partial p}{\partial x_i} dx_i + X_i dx_i \qquad (8),$$

since

$$\frac{\partial u_i}{\partial x_k} = \frac{\partial^2 \phi}{\partial x_i \partial x_k} = \frac{\partial u_k}{\partial x_i} \qquad (9).$$

This shows that for all contemporaneous variations

$$\frac{\partial\phi}{\partial t} + \tfrac{1}{2}u_k{}^2 = -\int\frac{dp}{\rho} + U + \text{constant} \qquad (10),$$

where

$$X_i = \frac{\partial U}{\partial x_i} \qquad (11);$$

$u_k{}^2$ is the square of the resultant velocity q. The constant of integration is not necessarily the same at all instants and therefore may be a function of the time. Hence we have the Bernoulli integral

$$\frac{\partial\phi}{\partial t} + \tfrac{1}{2}q^2 = -\int\frac{dp}{\rho} + U + F(t) \qquad (12).$$

The rate of change of mass within a given small parallelepiped $dx_1 dx_2 dx_3$ is equal and opposite to the rate of outflow; hence

$$\frac{\partial\rho}{\partial t}\,d\tau = -\frac{\partial}{\partial x_i}(\rho u_i)\,d\tau \qquad (13),$$

and we have the equation of continuity

$$\frac{\partial\rho}{\partial t} = -\frac{\partial}{\partial x_i}(\rho u_i) \qquad (14),$$

or

$$\frac{d\rho}{dt} = -\rho\frac{\partial u_i}{\partial x_i} \qquad (15).$$

Vectors with given Divergence and Curl. We sometimes have to find a vector u_i such that

$$\frac{\partial u_i}{\partial x_i} = \Delta \qquad (1),$$

$$\frac{\partial u_k}{\partial x_i} - \frac{\partial u_i}{\partial x_k} = \omega_{ik} \qquad (2),$$

where Δ is a given scalar and ω_{ik} a given antisymmetrical tensor. We want particular integrals of these equations.

Evidently if

$$u_i = \frac{\partial\phi}{\partial x_i} \qquad (3),$$

where ϕ is any scalar,

$$\nabla^2\phi = \Delta \qquad (4),$$

$$\frac{\partial u_k}{\partial x_i} - \frac{\partial u_i}{\partial x_k} = 0 \qquad (5),$$

and a solution of (4) is

$$\phi = -\frac{1}{4\pi}\iiint \frac{\Delta}{r}\,d\tau \qquad (6),$$

where ϕ is to be evaluated at x_i and Δ at ξ_i;

$$d\tau = d\xi_1 d\xi_2 d\xi_3 \qquad (7),$$

and r is the distance from ξ_i to x_i.

If
$$u_i = \frac{\partial F_m}{\partial x_k} - \frac{\partial F_k}{\partial x_m} = \epsilon_{ikm}\frac{\partial F_m}{\partial x_k} \qquad (8),$$

where F_i is a vector such that

$$\frac{\partial F_i}{\partial x_i} = 0 \qquad (9),$$

we have
$$\frac{\partial u_i}{\partial x_i} = \epsilon_{ikm}\frac{\partial^2 F_m}{\partial x_i \partial x_k} = 0 \qquad (10).$$

Then also

$$\begin{aligned}
\omega_m = \frac{\partial u_k}{\partial x_i} - \frac{\partial u_i}{\partial x_k} &= \epsilon_{ikm}\frac{\partial u_k}{\partial x_i} \\
&= \epsilon_{ikm}\frac{\partial}{\partial x_i}\epsilon_{kps}\frac{\partial F_s}{\partial x_p} \\
&= -\epsilon_{ikm}\epsilon_{pks}\frac{\partial^2 F_s}{\partial x_i \partial x_p} \\
&= (\delta_{is}\delta_{mp} - \delta_{ip}\delta_{ms})\frac{\partial^2 F_s}{\partial x_i \partial x_p} \\
&= \delta_{is}\frac{\partial^2 F_s}{\partial x_i \partial x_m} - \delta_{ip}\frac{\partial^2 F_m}{\partial x_i \partial x_p} \\
&= \frac{\partial}{\partial x_m}\frac{\partial F_i}{\partial x_i} - \frac{\partial^2 F_m}{\partial x_i^2} \\
&= -\nabla^2 F_m \qquad (11)
\end{aligned}$$

by (9).

Thus if
$$F_m = \frac{1}{4\pi}\iiint \frac{\omega_m}{r}\,d\tau \qquad (12$$

we shall have

$$\frac{\partial u_k}{\partial x_i} - \frac{\partial u_i}{\partial x_k} = \omega_{ik};\ \frac{\partial u_i}{\partial x_i} = 0 \qquad (13),$$

provided (9) is satisfied. But

$$\frac{\partial F_m}{\partial x_m} = \frac{1}{4\pi}\iiint \omega_m \frac{\partial}{\partial x_m}\left(\frac{1}{r}\right) d\tau$$

$$= -\frac{1}{4\pi}\iiint \omega_m \frac{\partial}{\partial \xi_m}\left(\frac{1}{r}\right) d\tau \qquad (14),$$

since ω_m is a function of ξ_i alone and r a function of $x_i - \xi_i$. Applying Green's Theorem to all space except a small sphere about x_i, we get

$$\frac{\partial F_m}{\partial x_m} = -\frac{1}{4\pi}\text{Lim}\iint \frac{l_m \omega_m}{r} dS + \frac{1}{4\pi}\text{Lim}\iiint \frac{1}{r}\frac{\partial \omega_m}{\partial \xi_m} d\tau \qquad (15),$$

since F_m has the form of a gravitation potential and $\partial F_m/\partial x_i$ that of a gravitational force. But the first integral vanishes in the limit when the sphere becomes very small, and the integrand in the second is zero provided the components u_i exist, for

$$\frac{\partial \omega_m}{\partial x_m} = \frac{\partial}{\partial x_m}\left(\frac{\partial u_k}{\partial x_i} - \frac{\partial u_i}{\partial x_k}\right) = \epsilon_{ikm}\frac{\partial^2 u_k}{\partial x_i \partial x_m} = 0 \qquad (16).$$

If then we are given a scalar Δ and a vector ω_m such that its divergence is zero, a solution of (1) and (2) is

$$u_i = \frac{\partial \phi}{\partial x_i} + \frac{\partial F_m}{\partial x_k} - \frac{\partial F_k}{\partial x_m} \qquad (17),$$

where ϕ and F_m are given by (6) and (12) and ikm are in cyclic order.

This analysis has two practical applications. In hydrodynamics u_i is the velocity and ω_m is twice the vorticity, denoted by $2\xi_m$ in Chapter IX. Here the divergence of the velocity and the vorticity may be given through all space, and the velocity (17) satisfies the conditions. If the actual velocity is v_i, we may put

$$v_i = u_i + u_i' \qquad (18),$$

and then

$$\frac{\partial u_i'}{\partial x_i} = 0; \quad \frac{\partial u_k'}{\partial x_i} - \frac{\partial u_i'}{\partial x_k} = 0 \qquad (19).$$

Thus u_i' is the gradient of a scalar ϕ' satisfying Laplace's equation. There is no such scalar that makes the velocity finite everywhere, including at an infinite distance, except a linear function of the coordinates. Hence u_i' is the same everywhere.

If there are solid boundaries or free surfaces at a finite distance u_i' may not be constant. If the region where vorticity is present does not extend to a boundary, there is no contribution to (14) from points outside this region; we therefore take (14) through a region large enough to contain the whole of the vorticity. Then the surface integral in (15) must also be taken over the boundary of this region, but still vanishes, and (17) is still a solution. But (17) may not satisfy the boundary conditions, and then we must add to u_i an irrotational solution chosen to make the whole velocity satisfy them.

In electromagnetism u_i may be the magnetic intensity and ω_m the electric current across unit surface in a plane of x_m constant.

In many cases u_i has no curl outside a limited region of very small cross-section. This is often true in the motion of a real fluid, when the region may be called a vortex filament, and in magnetism, when the region is a wire carrying an electric current. The former statement may be expressed also by saying that the motion is irrotational outside the vortex filament; the latter says that magnetic forces due to electric currents have a potential. In either case the integral

$$\Omega = \int_C u_i dx_i \qquad (20),$$

taken around a closed circuit, is zero if the circuit can be filled up by a surface not cutting the filament or the wire, and has a constant value for all circuits that cannot be so filled up. The two conditions are mutually exclusive, and therefore the critical region itself must be a closed circuit.

In each case Δ vanishes. Then $\phi = 0$, and the component F_m is given by

$$F_m = \frac{1}{4\pi}\iiint \frac{\omega_m}{r}\,d\tau \qquad (21)$$

taken through the critical region. But if we consider the contribution from an element between two planes separated by $d\xi_m$, and call the element of surface in a plane parallel to these dS, we have

$$d\tau = d\xi_m dS \qquad (22),$$

$$\iint \omega_m dS = \int u_i d\xi_i = \Omega \qquad (23),$$

where the line integral is taken around the boundary of the filament. Hence

$$F_m = \frac{\Omega}{4\pi}\int \frac{d\xi_m}{r} \qquad (24),$$

the integral being taken around the length of the filament. Also

$$\begin{aligned} u_i &= \frac{\Omega}{4\pi}\left\{\frac{\partial}{\partial x_k}\int \frac{d\xi_m}{r} - \frac{\partial}{\partial x_m}\int \frac{d\xi_k}{r}\right\} \\ &= \frac{\Omega}{4\pi}\,\epsilon_{ikm}\int \frac{\partial}{\partial x_k}\left(\frac{1}{r}\right) d\xi_m \\ &= \frac{\Omega}{4\pi}\int \epsilon_{ikm}\,\frac{(\xi_k - x_k)\,l_m}{r^3}\,ds \qquad (25), \end{aligned}$$

where ds is an element of length of the filament and l_m a direction cosine.

In hydrodynamics Ω is the circulation around the filament. In electromagnetism the unit current is such that if it flows in a circle of radius 1 cm. it produces magnetic force 2π at the centre. If we take the circle to be in the plane of x_3 constant, with its centre at the origin, we have

$$ds = d\theta;\ r = 1;\ x_1 = x_2 = x_3 = 0;$$

$$\xi_1 = \cos\theta,\ \xi_2 = \sin\theta,\ \xi_3 = 0;\ l_1 = -\sin\theta,\ l_2 = \cos\theta,\ l_3 = 0$$

and

$$\int \epsilon_{ikm} \frac{\xi_k l_m}{r^3} ds = 0 \text{ for } i = 1 \text{ or } 2$$

and

$$= 2\pi \text{ for } i = 3.$$

Also for unit current u_3, the magnetic force, $= 2\pi$.

Hence, in this case, by (25),

$$\Omega = 4\pi,$$

and in general, if the current is I, $\Omega = 4\pi I$, and

$$u_i = I \int \epsilon_{ikm} \frac{(\xi_k - x_k) l_m}{r^3} ds \qquad (26)$$

$$= I \int \epsilon_{ikm} \frac{\xi_k - x_k}{r^3} d\xi_m \qquad (27).$$

This may be transformed by Stokes's Theorem into an integral over a surface with the wire as its boundary; thus

$$u_i = I \iint l_m \epsilon_{mps} \frac{\partial}{\partial \xi_p} \epsilon_{iks} \frac{\xi_k - x_k}{r^3} dS$$

$$= - I \iint l_m \epsilon_{iks} \epsilon_{mps} \frac{\partial^2}{\partial \xi_k \partial \xi_p} \left(\frac{1}{r}\right) dS$$

$$= - I \iint l_m (\delta_{im} \delta_{kp} - \delta_{ip} \delta_{km}) \frac{\partial^2}{\partial \xi_k \partial \xi_p} \left(\frac{1}{r}\right) dS$$

$$= - I \iint l_m \left\{\delta_{im} \nabla^2 \left(\frac{1}{r}\right) - \frac{\partial^2}{\partial \xi_i \partial \xi_m} \left(\frac{1}{r}\right)\right\} dS$$

$$= - I \iint l_m \frac{\partial}{\partial \xi_m} \frac{\partial}{\partial x_i} \left(\frac{1}{r}\right) dS \qquad (28).$$

But $- l_m \dfrac{\partial^2}{\partial \xi_m \partial x_i} \left(\dfrac{1}{r}\right)$ is the magnetic force at x_i due to a doublet of unit strength at ξ_i with its axis in the direction l_m. Hence the force at any point is equivalent to that due to a distribution of doublets over the closing surface, with intensity I per unit area and directed normally to the surface. Such a distribution constitutes a *magnetic shell* of strength I.

It is customary to assert as a fundamental postulate that the magnetic force due to an electric current is equivalent to that of a magnetic shell, but I think this course undesirable. The magnetic shell does not exist in nature and direct experimental test is therefore impossible. Further, though it gives the same force, it does not give the same potential; the magnetic potential due to a magnetic shell is a single-valued function with a finite discontinuity at the shell, while that due to an electric current is a cyclic function with no discontinuity except at the wire itself.

Mutual Energy of Electric Circuit and Magnetic Field. A magnetic pole m at x_i is under a force mu_i; but

$$mu_i = -mI \frac{\partial}{\partial x_i} \iint l_m \frac{\partial}{\partial \xi_m} \left(\frac{1}{r}\right) dS \qquad (29)$$

and $-l_m \dfrac{\partial}{\partial \xi_m}\left(\dfrac{1}{r}\right)$ is the magnetic force normal to the surface at ξ_m due to the pole. Hence $-\iint l_m \dfrac{\partial}{\partial \xi_m}\left(\dfrac{1}{r}\right) dS$ is the total flux across the closing surface, or through the circuit, of the magnetic force due to the pole. Also if V is the magnetic potential due to the current,

$$u_i = -\frac{\partial V}{\partial x_i} \qquad (30),$$

and therefore the correct forces are given by taking

$$V = -I \iint N\,dS \qquad (31),$$

where N is the normal magnetic force due to a unit magnetic pole. This can now be generalized; we say that in general the mutual potential W of a current and a set of magnetic poles is

$$W = -I \iint N\,dS \qquad (32),$$

where N is now the normal magnetic force due to the whole of the rest of the field. This form may be further extended to express the mutual energy of two electric currents.

From this equation we can infer, as is done in the standard works of Jeans and Livens, that if $\iint N\,dS$ varies with the time, the variation generates an induced E.M.F. $-\frac{d}{dt}\iint N\,dS$ in the circuit.

If we have two circuits carrying currents I and J, their mutual influence is expressed by the statement that their mutual energy is of the form (32), where N is taken to be the force near ξ_i due to the current J; points on the latter circuit can be taken to be given by x_i. Then

$$W = IJM \qquad (33),$$

where

$$M = \iint l_i u_i \, dS \qquad (34),$$

and u_i is the magnetic force at ξ_i due to unit current in the second circuit. But if F_i is the vector potential due to such a current,

$$u_i = \epsilon_{ikm} \frac{\partial F_m}{\partial \xi_k} \qquad (35),$$

$$M = \iint \epsilon_{ikm} l_i \frac{\partial F_m}{\partial \xi_k} dS \qquad (36)$$

$$= \int F_m d\xi_m \qquad (37),$$

taken around the circuit. But

$$F_m = \int \frac{dx_m}{r} \qquad (38),$$

taken around the second circuit; and therefore

$$M = \iint \frac{dx_m d\xi_m}{r} \qquad (39),$$

taken around both circuits. This gives the required form for the coefficient of mutual induction of two circuits.

CHAPTER VII

ISOTROPIC TENSORS

A tensor is called isotropic if its components retain the same values however the axes are rotated. We have already obtained three examples, namely δ_{ik}, ϵ_{ikm}, and $\epsilon_{iks}\epsilon_{mps}$.

There are no isotropic tensors of the first order. For if u_i was such a tensor, let us give the axes a small rotation expressed by the antisymmetrical tensor c_{ik}. Then in the new system

$$u_i' = (\delta_{ij} - c_{ij})\, u_j = u_i - c_{ij} u_j \qquad (1),$$

and this can be equal to u_i only if

$$c_{ij} u_j = 0 \qquad (2)$$

for all admissible values of the c_{ij}. Thus

$$\left.\begin{aligned} c_{11}u_1 + c_{12}u_2 + c_{13}u_3 = 0 \\ c_{21}u_1 + c_{22}u_2 + c_{23}u_3 = 0 \\ c_{31}u_1 + c_{32}u_2 + c_{33}u_3 = 0 \end{aligned}\right\} \qquad (3).$$

But $c_{11} = c_{22} = c_{33} = 0$, while c_{12}, c_{23}, c_{31} are independent and equal and opposite to the components obtained by interchanging suffixes. Hence (3) can be satisfied only if

$$u_1 = u_2 = u_3 = 0 \qquad (4),$$

and therefore there is no isotropic tensor of the first order other than zero.

If u_{ik} is an isotropic tensor of the second order,

$$\begin{aligned} u'_{ik} &= (\delta_{ij} - c_{ij})(\delta_{kl} - c_{kl})\, u_{jl} \\ &= u_{ik} - c_{ij}\delta_{kl}u_{jl} - c_{kl}\delta_{ij}u_{jl} \\ &= u_{ik} - c_{ij}u_{jk} - c_{kl}u_{il} \qquad (1) \end{aligned}$$

to the first order, for all values of i and k. Hence

$$c_{ij}u_{jk} + c_{kj}u_{ij} = 0 \qquad (2).$$

If i and k are unequal, take $i = 1$, $k = 2$. Since $c_{11} = c_{22} = 0$ we have

$$c_{12} u_{22} + c_{13} u_{32} + c_{21} u_{11} + c_{23} u_{13} = 0 \qquad (3),$$

and therefore

$$u_{32} = u_{13} = 0; \; u_{11} = u_{22} \qquad (4).$$

By symmetry u_{ik} is therefore 0 if $i \neq k$, while $u_{11} = u_{22} = u_{33}$.

If i and k are both 1, we have

$$c_{12} u_{21} + c_{13} u_{31} + c_{12} u_{12} + c_{13} u_{13} = 0 \qquad (5),$$

which is satisfied since every term vanishes. Hence the only isotropic tensor of order 2 is a scalar multiple of δ_{ik}.

If u_{ikm} is an isotropic tensor of the third order,

$$u'_{ikm} = (\delta_{ij} - c_{ij})(\delta_{kl} - c_{kl})(\delta_{mn} - c_{mn}) u_{jln} \quad (1),$$

and therefore, for all values of i, k, m,

$$c_{ij} u_{jkm} + c_{kj} u_{ijm} + c_{mj} u_{ikj} = 0 \qquad (2).$$

Take $i = k = 1$. Then

$$c_{12} u_{21m} + c_{13} u_{31m} + c_{12} u_{12m} + c_{13} u_{13m} + c_{m1} u_{111} + c_{m2} u_{112} + c_{m3} u_{113} = 0 \qquad (3).$$

Now put $m = 2$ so that $c_{m2} = 0$. Then

$$\left.\begin{aligned} u_{212} + u_{122} &= u_{111} \\ u_{312} + u_{132} &= u_{112} \\ u_{113} &= 0 \end{aligned}\right\} \qquad (4).$$

From the last equation, and by symmetry, $u_{ikm} = 0$ if two of i, k, m are equal and the third unequal. Then by the first, u_{ikm} is also zero if all of i, k, m are equal; and the second shows that

$$u_{ikm} = - u_{kim}.$$

If in (3) we put $m = 1$, every term vanishes, so that (3) holds.

Now in (2) if i, k, m are all different, u_{jkm} is zero unless $j = i$, and then $c_{ij} = 0$. Hence (2) holds. It follows that the only isotropic tensors of order 3 are scalar multiples of ϵ_{ikm}.

If u_{ikmp} is an isotropic tensor of order 4, we have, similarly,

$$c_{ij}u_{jkmp} + c_{kj}u_{ijmp} + c_{mj}u_{ikjp} + c_{pj}u_{ikmj} = 0 \quad (1).$$

There are only three possible values for i, k, m, p, and therefore at least two of them must be equal. We may consider separately the cases where (a) two are equal and the other two unequal, (b) three equal, (c) two equal and the other two equal, (d) all four equal.

In case (a), take $i = k = 1$, $m = 2$, $p = 3$. Then

$$c_{12}u_{2123} + c_{13}u_{3123} + c_{12}u_{1223} + c_{13}u_{1323} + c_{21}u_{1113} + c_{23}u_{1133} + c_{31}u_{1121} + c_{32}u_{1122} = 0 \quad (2).$$

Hence, by the antisymmetrical property of c_{ik},

$$\left.\begin{aligned} u_{2123} + u_{1223} - u_{1113} &= 0 \\ u_{3123} + u_{1323} - u_{1121} &= 0 \end{aligned}\right\} \quad (3),$$

$$u_{1133} - u_{1122} = 0 \quad (4).$$

Other instances of case (a) can be obtained by interchanging suffixes that are not already equal, and by turning the axes so as to bring 3 into the position of 1, 1 into that of 2, and 2 into that of 3. Thus (4) gives

$$u_{1133} = u_{1122} = u_{2233} = u_{2211} = u_{3322} = u_{3311} \quad (5).$$

And also

$$u_{1313} = u_{1212} = u_{2323} = u_{2121} = u_{3232} = u_{3131} \quad (6),$$

$$u_{3113} = u_{2112} = u_{3223} = u_{1221} = u_{2332} = u_{1331} \quad (7).$$

In case (b), take $i = k = m = 1$, $p = 2$.

$$c_{12}u_{2112} + c_{13}u_{3112} + c_{12}u_{1212} + c_{13}u_{1312} + c_{12}u_{1122} + c_{13}u_{1132} + c_{21}u_{1111} + c_{23}u_{1113} = 0 \quad (8).$$

The last term shows that

$$u_{1113} = 0 \quad (9),$$

and therefore, by interchange of suffixes, all components of class (*b*) are zero. Also, from the coefficient of c_{13},

$$u_{3112} + u_{1312} + u_{1132} = 0 \qquad (10).$$

But in (3) the last term vanishes and we infer

$$u_{2123} + u_{1223} = 0 \qquad (11),$$

and therefore

$$u_{1312} + u_{3112} = 0 \qquad (12),$$

whence, by (10),

$$u_{1132} = 0 \qquad (13).$$

Thus all components of class (*a*) are also zero.

The coefficient of c_{12} in (8) gives

$$u_{1111} = u_{2112} + u_{1212} + u_{1122} \qquad (14),$$

so that the components of class (*d*) are expressible in terms of the three types of class (*c*).

No further information is got by transforming components of classes (*c*) and (*d*). Thus if $i = k = 1$, $m = p = 2$, replacing i or k by j will give a zero component unless j is equal to 1; and then the factor c_{ij} or c_{kj} is zero, and the relation holds automatically. Similar considerations apply if all of i, k, m, p are equal.

We may denote the components of type (5) by λ, those of type (6) by $\mu + \nu$, and those of type (7) by $\mu - \nu$. Then (14) gives

$$u_{1111} = u_{2222} = u_{3333} = \lambda + 2\mu \qquad (15).$$

There appear therefore to be three independent isotropic tensors of order 4, obtained by taking each of λ, μ, ν in turn equal to 1 and the others to zero.

In the λ tensor, $u_{ikmp} = 1$ if $i = k$ and $m = p$, and in all other cases is zero. It is therefore equivalent to $\delta_{ik}\delta_{mp}$, which is obviously a tensor of order 4, being the product of two tensors of order 2.

In the μ tensor, $u_{ikmp} = 1$ if $i = m$, $k = p$, or if $i = p$,

$k = m$, and $i \neq k$. If also $i = k$, the component is 2. Other components are zero. This can be written

$$u_{ikmp} = \delta_{im}\delta_{kp} + \delta_{ip}\delta_{km} \qquad (16),$$

and is obviously a tensor of order 4.

In the ν tensor, $u_{ikmp} = 1$ if $i = m$, $k = p$, and $= -1$ if $i = p$, $k = m$, and in all other cases is zero. If also $i = k$, u_{ikmp} is zero. In this case, therefore,

$$u_{ikmp} = \delta_{im}\delta_{kp} - \delta_{ip}\delta_{km} \qquad (17).$$

This can also be written

$$u_{ikmp} = \epsilon_{ijk}\epsilon_{mjp} \qquad (18),$$

for if $i = 1$, $k = 3$, $\epsilon_{ijk} = 0$ unless $j = 2$ and then $= 1$. But then $\epsilon_{mjp} = 1$ if $m = 1$, $p = 3$, -1 if $m = 3$, $p = 1$, and otherwise $= 0$. Thus

$$u_{1313} = 1, \; u_{1331} = -1 \qquad (19),$$

with corresponding values for the other components. Evidently (17) and (18) represent a tensor of order 4. It has already appeared in Chapters I and VI.

The general isotropic tensor of order 4 is therefore

$$\lambda\delta_{ik}\delta_{mp} + \mu(\delta_{im}\delta_{kp} + \delta_{ip}\delta_{km}) + \nu(\delta_{im}\delta_{kp} - \delta_{ip}\delta_{km}) \qquad (20),$$

where λ, μ, ν are scalars.

EXAMPLE

Prove that

$$\delta_{ik}\delta_{mp}w_{ik} = \delta_{mp}w_{ii},$$

$$(\delta_{im}\delta_{kp} + \delta_{ip}\delta_{km})\,w_{ik} = w_{mp} + w_{pm},$$

$$(\delta_{im}\delta_{kp} - \delta_{ip}\delta_{km})\,w_{ik} = w_{mp} - w_{pm}.$$

CHAPTER VIII

ELASTICITY

In an elastic solid, as in a fluid, the distance between any two particles of the body usually varies with the time. The body, however, has an equilibrium configuration that could persist if the external forces were zero or constant. We may take this as a standard of reference. If a particle actually at x_i would be at $x_i - u_i$ in the standard configuration, we call u_i the displacement at x_i; in practice the squares of the u_i can usually be neglected. Evidently u_i is a vector.

At a point $x_i + y_i$, where y_i is small, the displacement is $u_i + v_i$, where

$$v_i = \frac{\partial u_i}{\partial x_k} y_k \tag{1}$$

$$= (e_{ik} - \xi_{ik})\, y_k \tag{2},$$

where e_{ik} and ξ_{ik} are the symmetrical and antisymmetrical tensors

$$e_{ik} = \tfrac{1}{2}\left(\frac{\partial u_i}{\partial x_k} + \frac{\partial u_k}{\partial x_i}\right); \quad \xi_{ik} = \tfrac{1}{2}\left(\frac{\partial u_k}{\partial x_i} - \frac{\partial u_i}{\partial x_k}\right) \tag{3}.$$

If e_{ik} is zero at x_i, the displacement has the same form as that due to a general displacement u_i together with a rotation expressed by the tensor ξ_{ik}. Also, if e_{ik} is everywhere zero,

$$\begin{aligned}\frac{\partial \xi_{ik}}{\partial x_m} &= \tfrac{1}{2}\frac{\partial}{\partial x_m}\left(\frac{\partial u_k}{\partial x_i} - \frac{\partial u_i}{\partial x_k}\right)\\ &= \tfrac{1}{2}\frac{\partial}{\partial x_i}\left(\frac{\partial u_k}{\partial x_m} + \frac{\partial u_m}{\partial x_k}\right) - \tfrac{1}{2}\frac{\partial}{\partial x_k}\left(\frac{\partial u_i}{\partial x_m} + \frac{\partial u_m}{\partial x_i}\right)\\ &= \frac{\partial e_{km}}{\partial x_i} - \frac{\partial e_{im}}{\partial x_k} = 0 \end{aligned} \tag{4}.$$

Hence the ξ_{ik} are constants and the rotation is the same everywhere. The vanishing of the e_{ik} is therefore the necessary and sufficient condition for a pure rotation.

Now consider the part of v_i due to the e_{ik}. If we consider the quadric surface

$$e_{ik}y_iy_k = r^2 \tag{5},$$

where r is a constant chosen so as to make the surface pass through y_i, the normal at y_i has direction cosines proportional to $e_{ik}y_k$ and therefore to v_i. The displacement due to e_{ik} is therefore parallel to the normal at y_i to this quadric. This coincides with the direction of the radius vector to y_i if y_i is on a principal axis. There are therefore three directions such that the relative displacement due to the e_{ik} is in the direction of y_i, and these directions are mutually perpendicular. If we take new axes ξ_1, ξ_2, ξ_3 along them, the quadric reduces to

$$e_{11}'\xi_1^2 + e_{22}'\xi_2^2 + e_{33}'\xi_3^2 = r^2 \tag{6},$$

and all terms e_{jl}' with $j \neq l$ are zero. The displacement in the ξ_1 direction is now $e_{11}'\xi_1$, so that all distances in this direction are increased in the ratio 1 to $1 + e_{11}'$. The displacement due to the e_{ik} is then the resultant of three homogeneous strains parallel to three orthogonal axes.

We see therefore that the displacement in any small neighbourhood can be represented as the combination of a rotation with three extensions at right angles. The latter express the changes of size and shape of an element of the solid. For this reason e_{ik} is called the strain tensor and ξ_{ik} the rotation tensor. Evidently e_{ik} has six independent components. For we can have

$$v_1 = ey_1,\ v_2 = v_3 = 0 \tag{7},$$

making all the e_{ik} zero except $e_{11} = e$; similarly e_{22} and e_{33} may exist independently of the others. Also, if

$$v_1 = 0,\ v_2 = ez,\ v_3 = ey \tag{8},$$

all the e_{ik} vanish except $e_{23} = e_{32} = e$. Similarly e_{31} and e_{12} can be assigned independently.

In an elastic solid the internal force across an element of surface is in general inclined to the surface. If the area of the element is dS, the force across it must be specified by three components of the form $p_{ni}dS$ parallel to the axes; where n may be regarded as indicating the normal to the surface. If we consider a small parallelepiped with edges dx_1, dx_2, dx_3, centred at (x_1, x_2, x_3), the force across the face of area dx_2dx_3 centred at $x_1 + \frac{1}{2}dx_1$ is $(p_{11}, p_{12}, p_{13})\,dx_2dx_3$, where the p_{1i} are evaluated at $x_1 + \frac{1}{2}dx_1$. The force on the opposite face is $-(p_{11}, p_{12}, p_{13})\,dx_2dx_3$ evaluated at $x_1 - \frac{1}{2}dx_1$, and the total is $\frac{\partial p_{1i}}{\partial x_1}dx_1dx_2dx_3$. In general the force in the x_i direction due to the stress across the faces of x_k constant is $\frac{\partial p_{ki}}{\partial x_k}dx_1dx_2dx_3$, and we take account of all faces by using the summation convention. If the acceleration of the element has components f_i and the density is ρ, the mass is $\rho\,dx_1dx_2dx_3$; while if the bodily force acting has components X_i per unit mass, the equations of motion are

$$\rho f_i = \frac{\partial p_{ki}}{\partial x_k} + \rho X_i \qquad (9).$$

The system of quantities p_{ik} constitutes a symmetrical tensor of the second order. To prove this, consider first a plane whose normal has direction cosines a_{ij}, intersecting lines through x_i parallel to the coordinate axes at short distances from x_i; thus a small tetrahedron is formed, whose sides are of order l, say. Let the area of the sloping face be dS; then those of the others are $a_{ij}dS$. Consider now the rate of change of momentum of the matter within this tetrahedron. Evidently the contributions from the acceleration and X_i are of the order of the volume, that is, of l^3. The force across dS has magnitude $p_{ji}dS$. That across the face of x_k constant is $-p_{ki}$ times the area of the face, that is, $-p_{ki}a_{kj}dS$. But dS is of order l^2. Hence we have

$$(p_{ji} - a_{kj}p_{ki})\,O\,(l^2) = O\,(l^3) \qquad (10),$$

and hence if l is indefinitely small we have for the stress across a plane normal to a_{ij} at x_i,

$$p_{ji} = a_{kj}p_{ki} \qquad (11).$$

Now consider three perpendicular directions with direction cosines a_{ij} ($j = 1, 2, 3$). The force per unit area across a plane perpendicular to one of these axes, in the direction of x_i, is given by p_{ji}. Resolving this along the direction of one of the new axes x_l we have, therefore,

$$p_{jl} = a_{kj}a_{il}p_{ki} = a_{ij}a_{kl}p_{ik} \qquad (12)$$

by interchanging i and k; which is precisely the law of transformation of a second order tensor.

Consider again a small parallelepiped centred at x_i, with edges parallel to the coordinate axes, and form the equation of moments about its centre. The contributions from f_k and X_k are of order l^4 at most, where the edges have lengths of order l. The moment about an axis parallel to x_3 of the stress p_{21} in the face $x_2 =$ constant and parallel to x_1 is the product of p_{21} into the area of the face and the distance of the face from the centre; that is, to order l^3, $\frac{1}{2}p_{21}\,dx_1\,dx_2\,dx_3$. The opposite face makes an equal contribution. The stress p_{12} in the face $x_2 =$ constant and parallel to x_1 tends to turn in the opposite direction. We have, therefore,

$$(p_{21} - p_{12})\,dx_1\,dx_2\,dx_3 = O\,(l^4) \qquad (13),$$

and therefore when we make l indefinitely small we must have

$$p_{21} = p_{12} \qquad (14),$$

and in general

$$p_{ik} = p_{ki} \qquad (15)$$

so that the tensor p_{ik} is symmetrical.

Now consider the energy interchange between the small parallelepiped and its surroundings. The stresses across the face of area $dx_2 dx_3$ centred at $x_1 + \frac{1}{2}dx_1$ are doing work on the element at a rate $(p_{k1}\dot{u}_k)\, dx_2 dx_3$, and the contribution from the two opposite faces is

$$\frac{\partial}{\partial x_1}(p_{k1}\dot{u}_k)\, dx_1 dx_2 dx_3.$$

Thus in all the stresses are doing work at a rate

$$\frac{\partial}{\partial x_k}(p_{ki}\dot{u}_i)\, d\tau.$$

The external forces are doing work at a rate $\rho X_i \dot{u}_i d\tau$. The kinetic energy of the element is $\frac{1}{2}\rho \dot{u}_i{}^2 d\tau$, and is increasing at a rate $\rho \dot{u}_i f_i d\tau$. (We consider the actual specimen of matter occupying the element of volume $d\tau$ at time t; thus its mass is $\rho d\tau$, and if we keep to the same piece of *matter* at time $t + dt$ the mass is unaltered. If we considered the variation of energy within a given element of volume we should have to allow for the variation of ρ and the fact that the matter moving out of the element is taking its energy with it.) The rate of performance of work on the element therefore exceeds the rate of increase of kinetic energy by

$$\left\{\frac{\partial}{\partial x_k}(p_{ki}\dot{u}_i) + \rho X_i \dot{u}_i - \rho \dot{u}_i f_i\right\} d\tau$$

$$= \left\{\frac{\partial}{\partial x_k}(p_{ki}\dot{u}_i) + \rho X_i \dot{u}_i - \dot{u}_i\left(\frac{\partial p_{ki}}{\partial x_k} + \rho X_i\right)\right\} d\tau,$$

by the equations of motion,

$$= \left(p_{ki}\frac{\partial \dot{u}_i}{\partial x_k}\right) d\tau \qquad (16).$$

This work is stored up as internal energy of the element of volume. Evidently from its form it is a scalar.

In any elastic solid the internal energy is a definite function of the state of the solid. In any change of state the increase of internal energy therefore depends only on the initial and final states and not on the method of passage from one to the other. Now we have seen that six of the e_{ik} are independent, and if the element acquires displacements δu_i in time δt the corresponding increase of internal energy is $p_{ki}\,\delta\left(\frac{\partial u_i}{\partial x_k}\right) d\tau$. There is an apparent asymmetry according as i and k are equal or unequal. Thus $p_{11}\,d\tau$ has coefficient $\delta\,(\partial u_1/\partial x_1) = \delta e_{11}$, but $i = 1$, $k = 2$ contributes $p_{21}\,\delta\,(\partial u_1/\partial x_2)$, and $i = 2$, $k = 1$ contributes $p_{12}\,\delta\,(\partial u_2/\partial x_1)$, the two together giving $2p_{12}\,\delta e_{12}$. But this is the same as $p_{12}\,\delta e_{12} + p_{21}\,\delta e_{21}$, and the whole contribution from the changes of strain is $p_{ik}\,\delta e_{ik}\,d\tau$. Also during the process an amount of heat $\delta Q\,d\tau$ may be absorbed. If then $E d\tau$ is the internal energy of the element,

$$\delta E = p_{ik}\,\delta e_{ik} + \delta Q \qquad (17).$$

Since E is a definite function of the state of the system, and six of the δe_{ik} are independent and determine the other three, δE depends on the changes of the temperature and of the six independent e_{ik} and has a definite value in whatever order these changes come about. But

$$\left(\frac{\partial E}{\partial e_{11}}\right)_{\delta Q=0} = p_{11}; \quad \left(\frac{\partial E}{\partial e_{12}}\right)_{\delta Q=0} = 2p_{12} \qquad (18).$$

If the absolute temperature is θ, and a certain amount of heat δQ is absorbed without change of any linear dimension, the rise of temperature is related to δQ by the rule

$$\delta Q = \rho c\,\delta\theta \qquad (19),$$

where c is the specific heat at constant strain. If there is also a change of strain, since δQ and $\delta\theta$ are scalars, we must have

$$\delta Q = q_{ik}\,\delta e_{ik} + \rho c\,\delta\theta \qquad (20),$$

where the q_{ik} constitute a tensor of the second order. But δE

and $\delta Q/\theta$ are perfect differentials. Hence if we replace the six independent e_{ik} by e_r we can write

$$\delta E = (p_r + q_r)\,\delta e_r + \rho c\,\delta\theta \tag{21}$$

$$\frac{\delta Q}{\theta} = \frac{q_r}{\theta}\,\delta e_r + \frac{\rho c}{\theta}\,\delta\theta \tag{22}$$

and

$$\frac{\partial}{\partial e_r}(p_s + q_s) = \frac{\partial}{\partial e_s}(p_r + q_r);\quad \frac{\partial}{\partial\theta}(p_r + q_r) = \frac{\partial}{\partial e_r}(\rho c) \tag{23}$$

$$\frac{\partial}{\partial e_r}\left(\frac{q_s}{\theta}\right) = \frac{\partial}{\partial e_s}\left(\frac{q_r}{\theta}\right);\quad \frac{\partial}{\partial\theta}\left(\frac{q_r}{\theta}\right) = \frac{\partial}{\partial e_r}\left(\frac{\rho c}{\theta}\right) \tag{24}$$

It follows at once that if θ is kept constant $\Sigma p_r\,\delta e_r$ and $\Sigma q_r\,\delta e_r$ are perfect differentials. Also

$$\frac{\partial}{\partial\theta}(p_r + q_r) = \theta\,\frac{\partial}{\partial e_r}\left(\frac{\rho c}{\theta}\right) = \theta\,\frac{\partial}{\partial\theta}\left(\frac{q_r}{\theta}\right) = \frac{\partial q_r}{\partial\theta} - \frac{q_r}{\theta} \tag{25}$$

and therefore

$$q_r = -\,\theta\,\frac{\partial p_r}{\partial\theta} \tag{26}$$

If $\delta\theta = 0$ we can write

$$\Sigma p_r\,\delta e_r = \delta W \tag{27}$$

where

$$2W = c_0 + 2c_r e_r + c_{rs} e_r e_s + O(e^3) \tag{28}$$

The c_0, c_r, c_{rs} may be functions of θ. Then if we retain only terms in W up to order e^2,

$$p_r = c_r + c_{rs} e_s \tag{29}$$

The c_r represent the stresses that would remain if the strains e_r were removed without change of temperature. In most practical cases the original state is one of uniform temperature and no stress, so that $c_r = 0$. If there is a rise of temperature θ' under no stress, an element will acquire displacements

$$v_i = a_{ik}\,\theta' y_k \tag{30}$$

where α_{ik} is a second order tensor expressing the thermal expansion. Thus

$$e_{ik} = \tfrac{1}{2}(\alpha_{ik} + \alpha_{ki})\,\theta' = \beta_{ik}\theta' \qquad (31),$$

where β_{ik} is a symmetrical tensor; and

$$p_{ik} = c_{ik} + c_{ik,mp}\beta_{mp}\theta' \qquad (32),$$

where $c_{ik,mp}$ is a fourth order tensor.

But by hypothesis this deformation takes place under no stress and therefore $p_{ik} = 0$. This determines c_{ik}, and our formula for the stress is

$$p_{ik} = c_{ik,mp}(e_{mp} - \beta_{mp}\theta') \qquad (33).$$

The coefficient $c_{ik,mp}$ is the coefficient of $e_{ik}e_{mp}$ in W. Since there are six independent e_r there are twenty-one possible terms in a quadratic form $c_{rs}e_r e_s$, and therefore there are twenty-one coefficients $c_{ik,mp}$. They clearly form a tensor of order 4; such a tensor in general would have eighty-one components, but this satisfies the symmetry relations that it is unaltered if we interchange i and k, or m and p, or i and k together with m and p together.

From (26) and (33),

$$q_{ik} = -\theta e_{mp}\frac{\partial}{\partial\theta}(c_{ik,mp}) + \theta\frac{\partial}{\partial\theta}(c_{ik,mp}\beta_{mp}\theta')$$

$$= -\theta e_{mp}\frac{\partial}{\partial\theta}(c_{ik,mp}) + \theta c_{ik,mp}\beta_{mp} \qquad (34)$$

if θ' is small. The second term does not involve the e_{ik}; the first is small of the first order in the e_{ik}.

Many solids are isotropic; that is, they have the same properties in all directions. This applies to vitreous (glassy) solids and to mixtures of crystals oriented at random. In that case a uniform rise of temperature in an element gives an equal expansion in all directions and

$$v_i = \alpha\theta' y_i \qquad (35)$$

simply; then α is the coefficient of linear expansion and

$$\beta_{ik} = \alpha\delta_{ik} \tag{36}$$

The second order terms in W_2 constitute a scalar; and we have

$$\frac{\partial p_{ik}}{\partial e_{mp}} = c_{ik,mp} \tag{37}$$

a tensor of order 4. If it is isotropic it must be of the form (20) of Chapter VII. Then the linear terms in p_{ik} give

$$\begin{aligned} p_{ik} &= c_{ik,mp} e_{mp} \qquad (38) \\ &= \lambda \delta_{ik} \delta_{mp} e_{mp} + \mu (\delta_{im} \delta_{kp} + \delta_{ip} \delta_{km}) e_{mp} \\ &\qquad + \nu (\delta_{im} \delta_{kp} - \delta_{ip} \delta_{km}) e_{mp} \\ &= \lambda \delta_{ik} e_{mm} + \mu (\delta_{im} e_{mk} + \delta_{ip} e_{kp}) + \nu (\delta_{im} e_{mk} - \delta_{ip} e_{kp}) \\ &= \lambda \delta_{ik} e_{mm} + \mu (e_{ik} + e_{ki}) + \nu (e_{ik} - e_{ki}) \\ &= \lambda \delta_{ik} e_{mm} + 2\mu e_{ik} \qquad (39), \end{aligned}$$

the last term vanishing since e_{ik} is symmetrical.

Then

$$\begin{aligned} 2W_2 &= p_{ik} e_{ik} \qquad (40) \\ &= \lambda e_{ii} e_{mm} + 2\mu e_{ik} e_{ik} \\ &= \lambda \Delta^2 + 2\mu (e_{11}{}^2 + e_{22}{}^2 + e_{33}{}^2 + 2e_{23}{}^2 + 2e_{31}{}^2 + 2e_{12}{}^2) \qquad (41), \end{aligned}$$

where $$\Delta = e_{ii} = \partial u_i / \partial x_i \tag{42}$$

The scalars λ and μ represent properties of the material. Both are positive.

We can also write

$$\begin{aligned} 2W_2 = (\lambda + 2\mu) \Delta^2 + 4\mu (e_{23}{}^2 + e_{31}{}^2 + e_{12}{}^2 \\ - e_{22} e_{33} - e_{33} e_{11} - e_{11} e_{22}) \qquad (43). \end{aligned}$$

This appears to differ from the form in Love's *Elasticity*, 1906, p. 100, but the present e_{ik} differ from Love's strain components. My e_{11} is the same as his, namely $\partial u_1/\partial x_1$; but my e_{23} is only half his, so that his assemblage of strain components is not a tensor.

If all the e_{ii} were equal to one another and therefore to $\frac{1}{3}\Delta$, we should have

$$2W_2 = (\lambda + \tfrac{2}{3}\mu)\,\Delta^2 \qquad (44),$$

$$p_{ik} = (\lambda + \tfrac{2}{3}\mu)\,\Delta \quad (i = k); \qquad p_{ik} = 0 \quad (i \neq k) \qquad (45).$$

In general we write

$$\lambda + \tfrac{2}{3}\mu = k \qquad (46),$$

and call k the bulk-modulus.

$$\begin{aligned}
2W_2 = {} & (\lambda + \tfrac{2}{3}\mu)\,\Delta^2 \\
& + 2\mu\,(e_{11}{}^2 + e_{22}{}^2 + e_{33}{}^2 + 2e_{23}{}^2 + 2e_{31}{}^2 + 2e_{12}{}^2) \\
& - \tfrac{2}{3}\mu\,(e_{11}{}^2 + e_{22}{}^2 + e_{33}{}^2 + 2e_{22}e_{33} + 2e_{33}e_{11} + 2e_{11}e_{22}) \\
= {} & (\lambda + \tfrac{2}{3}\mu)\,\Delta^2 \\
& + \tfrac{4}{3}\mu\,(e_{11}{}^2 + e_{22}{}^2 + e_{33}{}^2 - e_{22}e_{33} - e_{33}e_{11} - e_{11}e_{22}) \\
& + 4\mu\,(e_{23}{}^2 + e_{31}{}^2 + e_{12}{}^2) \\
= {} & (\lambda + \tfrac{2}{3}\mu)\,\Delta^2 \\
& + \tfrac{2}{3}\mu\,\{(e_{22} - e_{33})^2 + (e_{33} - e_{11})^2 + (e_{11} - e_{22})^2 \\
& \qquad + 6e_{23}{}^2 + 6e_{31}{}^2 + 6e_{12}{}^2\} \qquad (47).
\end{aligned}$$

The coefficient of μ vanishes if and only if the strain is a symmetrical expansion, and may therefore be called the distortional strain energy.

If we allow for variations in temperature,

$$\begin{aligned}
p_{ik} &= \lambda\,(\Delta - 3\alpha\theta')\,\delta_{ik} + 2\mu\,(e_{ik} - \alpha\theta'\delta_{ik}) \\
&= \{\lambda\Delta - (3\lambda + 2\mu)\,\alpha\theta'\}\,\delta_{ik} + 2\mu\,e_{ik} \qquad (48)
\end{aligned}$$

and $$q_{ik} = -\,\theta\,\frac{\partial p_{ik}}{\partial\theta} \qquad (49).$$

Every term in p_{ik} is of the first order in the displacements; but $\partial\theta'/\partial\theta = 1$ and therefore gives rise to a constant term. This term is

$$q_{ik} = \theta\,.\,3k\alpha\,\delta_{ik} \qquad (50).$$

If $\delta Q = 0$, so that no heat is lost or gained by conduction,

$$\rho c\,\delta\theta = -\,q_{ik}\,\delta e_{ik} = -\,3k\alpha\theta\,\delta_{ik}\,e_{ik}$$
$$= -\,3k\alpha\theta\,\delta\Delta \qquad (51),$$

and therefore, if the strain takes place adiabatically,

$$\theta' = -\,\frac{3k\alpha\theta}{\rho c}\,\Delta \qquad (52),$$

and

$$p_{ik} = \left(\lambda + \frac{9k^2\alpha^2\theta}{\rho c}\right)\delta_{ik}\Delta + 2\mu e_{ik} \qquad (53).$$

Thus in an adiabatic disturbance the constant λ is increased above its value for a standard disturbance to λ', where

$$\lambda' = \lambda + \frac{9k^2\alpha^2\theta}{\rho c} \qquad (54),$$

while μ is unaltered. The bulk-modulus k is therefore also increased to k', where

$$k' = k + \frac{9k^2\alpha^2\theta}{\rho c} \qquad (55).$$

In a simple thermal expansion at zero stress the absorption of heat δQ is equal to $\rho c_p\,\delta\theta$, where c_p is called the specific heat at zero stress, and is the specific heat measured in ordinary experiments. Then

$$\rho c_p\,\delta\theta = \delta Q = \rho c\,\delta\theta + q_{ik}\,\delta e_{ik}$$
$$= \rho c\,\delta\theta + 3k\alpha\theta\,\delta_{ik}\,\delta e_{ik}$$
$$= \rho c\,\delta\theta + 3k\alpha\theta\,.\,3\alpha\,\delta\theta.$$

Thus

$$c_p = c\left(1 + \frac{9k\alpha^2\theta}{\rho c}\right) = \frac{k'c}{k} \qquad (56).$$

The equations of motion at constant temperature, if the properties λ and μ are uniform, can be written

$$\rho f_i = \frac{\partial}{\partial x_k}\,(\lambda\delta_{ik}\Delta + 2\mu\,e_{ik}) + \rho X_i \qquad (57)$$
$$= \frac{\partial}{\partial x_i}\,(\lambda\Delta) + \mu\,\frac{\partial}{\partial x_k}\left(\frac{\partial u_k}{\partial x_i} + \frac{\partial u_i}{\partial x_k}\right) + \rho X_i$$
$$= \frac{\partial}{\partial x_i}\,\{(\lambda + \mu)\,\Delta\} + \mu\nabla^2 u_i + \rho X_i \qquad (58).$$

If the changes are adiabatic λ must be replaced by λ'.

If there is any heat conduction, the absorption of heat per unit time by the element of volume $d\tau$ is

$$\frac{\partial}{\partial x_i}\left(K\frac{\partial \theta}{\partial x_i}\right) d\tau,$$

where K is the thermal conductivity. Then the equation of heat conduction is

$$\rho c\frac{\partial \theta}{\partial t} + q_{ik}\frac{\partial e_{ik}}{\partial t} = K\nabla^2\theta \qquad (59).$$

If we write

$$e_{ik} = \alpha\delta_{ik}\theta' + e_{ik}' \qquad (60),$$

so that e_{ik}' is the strain due to the stresses,

$$q_{ik}\frac{\partial e_{ik}}{\partial t} = 9k\alpha^2\theta\frac{\partial \theta}{\partial t} + 3k\alpha\,\theta\Delta' \qquad (61),$$

and the equation becomes

$$\rho c_p\frac{\partial \theta}{\partial t} + 3k\alpha\theta\frac{\partial \Delta'}{\partial t} = K\nabla^2\theta \qquad (62).$$

The element $d\tau$ originally had volume

$$\frac{\partial\,(x_1 - u_1,\, x_2 - u_2,\, x_3 - u_3)}{\partial\,(x_1,\, x_2,\, x_3)}\,d\tau$$

$$= \left\{1 - \left(\frac{\partial u_1}{\partial x_1} + \frac{\partial u_2}{\partial x_2} + \frac{\partial u_3}{\partial x_3}\right) + O\,(e_{ik})^2\right\} d\tau \qquad (63),$$

so that its density was $\rho\,(1 + \Delta)$. Hence

$$\frac{d\Delta}{dt} = -\frac{1}{\rho}\frac{d\rho}{dt} \qquad (64).$$

This is the equation of continuity.

CHAPTER IX

HYDRODYNAMICS

In comparison with a typical elastic solid, a real fluid shows a great resemblance and a fundamental difference. The force per unit area across an element of surface parallel to a coordinate plane again constitutes a symmetrical tensor of order 2, and for the same reasons. The equations of motion are still (9) of Chapter VIII, and also the rate of performance of work on an element of volume $d\tau$ still has the form (16). The difference is that the internal energy in a fluid does not depend directly on how much it has been deformed. However the fluid is moved about and stirred up, provided it returns to its initial position, density, and temperature, the initial and final internal energies are equal. The deformation, however great it may be, makes no contribution; the stresses do work on each element, and thereby supply energy, but this is removed in restoring the original temperature. If energy of deformation existed in a fluid, all particles of it would have a tendency to return spontaneously to their original positions when stresses are removed, and they have none. Accordingly, while the rate of performance of work on the element of volume $d\tau$ in time dt is still $p_{ki}\partial\dot{u}_i/\partial x_k d\tau$, where the symbols have the same meanings as in elasticity, we can no longer assert from this the equation (18) of Chapter VIII, because the change of internal energy is not determinate when the changes of the e_{ik} are given. The e_{ik} may be as great as we like, but the energy does not increase indefinitely apart from changes in density and temperature; and the fluid moves in the same way under the same external forces whatever its previous history. We may say that an elastic solid has a memory; a fluid has none.

The stresses in a fluid are related, not to the total deformation, but to the rate of increase of deformation. In the former notation these have components

$$\frac{\partial e_{ik}}{\partial t} \text{ or } \tfrac{1}{2}\left(\frac{\partial \dot{u}_i}{\partial x_k} + \frac{\partial \dot{u}_k}{\partial x_i}\right).$$

The velocities now appear, instead of the displacements from an initial configuration, and we now denote the *velocities*, instead of the *displacements*, by u_i.

We also write

$$e_{ik} = \tfrac{1}{2}\left(\frac{\partial u_i}{\partial x_k} + \frac{\partial u_k}{\partial x_i}\right); \quad \xi_{ik} = \tfrac{1}{2}\left(\frac{\partial u_k}{\partial x_i} - \frac{\partial u_i}{\partial x_k}\right) \qquad (1),$$

so that e_{ik} now denotes the *rate of increase* of strain and ξ_{ik} the local angular velocity, usually called the *vorticity*. Now we say that p_{ik} is linearly related to the new e_{ik}; and therefore

$$p_{ik} = P_{ik} + c_{ik,\,mp}e_{mp} \qquad (2),$$

where P_{ik} is a symmetrical tensor of order 2 that possibly does not vanish with the e_{ik}, and $c_{ik,\,mp}$ is a tensor of order 4. Further, in a fluid at rest the stress is an isotropic tensor. Hence P_{ik} is isotropic and must be of the form $-p\delta_{ik}$, where p is a scalar, and $c_{ik,\,mp}$ must be of the form (20) of Chapter VII. On carrying out the summation with regard to m and p the term in ν disappears and we have

$$p_{ik} = -p\delta_{ik} + \lambda\delta_{ik}e_{mm} + 2\mu e_{ik} \qquad (3),$$

where λ and μ are scalars, at present undetermined. Evidently p and λ are not both necessary to preserve the form, and we may introduce the further convention

$$p_{ii} = -3p \qquad (4),$$

so that $-p$ is the mean of the three p_{ik} with equal suffixes. This gives

$$-3p = -3p + 3\lambda e_{mm} + 2\mu e_{ii} \qquad (5),$$

and therefore $$\lambda = -\tfrac{2}{3}\mu \qquad (6).$$

If we consider the internal energy as a function of the density ρ and temperature θ alone, we may consider the energy change in a symmetrical expansion. If the density increases by $\delta\rho$, there is a contraction in all dimensions in the ratio $\frac{1}{3}\delta\rho/\rho$ and the stresses do work

$$-\tfrac{1}{3}p_{ii}\frac{\delta\rho}{\rho}=p\frac{\delta\rho}{\rho} \quad (7).$$

At the same time there may an absorption or generation of heat; then the energy change is

$$\delta E = p\frac{\delta\rho}{\rho}+\delta Q \quad (8),$$

and the heat absorbed may be written

$$\delta Q = M\delta\rho+\rho c\delta\theta \quad (9),$$

where M is unknown and c is the specific heat at constant volume. Then

$$\delta E = \left(\frac{p}{\rho}+M\right)\delta\rho+\rho c\delta\theta \quad (10),$$

$$\frac{\delta Q}{\theta}=\frac{M}{\theta}\delta\rho+\frac{\rho c}{\theta}\delta\theta \quad (11),$$

and the condition that these quantities are perfect differentials gives

$$M=-\theta\frac{\partial}{\partial\theta}\left(\frac{p}{\rho}\right) \quad (12),$$

$$\frac{\partial}{\partial\rho}(\rho c)=-\theta\frac{\partial^2}{\partial\theta^2}\left(\frac{p}{\rho}\right) \quad (13).$$

It appears also that p must be a function of ρ and θ alone, for a given material; it does not involve $\partial\rho/\partial t$.* We may call p the *pressure*.

* This amounts to saying that there is no dissipation of energy in a symmetrical compression or expansion. This is true in a gas on the older kinetic theory; but Enskog has shown (*Kungl. Svenska Akad. Handlingar*, 63, no. 4, 1922, p. 18) that p can with greater accuracy be given by

$$p=\frac{R}{M}\rho\theta+\frac{\eta}{\rho}\frac{d\rho}{dt},$$

where η is a "second" coefficient of viscosity. But η/μ is only of the order of the square of the ratio of the volume of the molecules themselves to the whole volume of the gas.

In a liquid the coefficient of $d\rho/dt$, if it exists, is within the experimental error; an analogous statement is true for imperfectly elastic solids.

The stress components may now be written

$$p_{ik} = -(p + \tfrac{2}{3}\mu e_{mm})\,\delta_{ik} + 2\mu e_{ik} \qquad (14),$$

where p is a function of the density and temperature and μ expresses a property of the fluid. We call μ the *coefficient of viscosity*. For a uniform fluid we have the equations of motion

$$\begin{aligned}\rho f_i &= \frac{\partial}{\partial x_k}\left\{-(p + \tfrac{2}{3}\mu e_{mm})\,\delta_{ik} + \mu\left(\frac{\partial u_i}{\partial x_k} + \frac{\partial u_k}{\partial x_i}\right)\right\} \\ &= -\frac{\partial}{\partial x_i}(p + \tfrac{2}{3}\mu e_{mm}) + \mu\nabla^2 u_i + \mu\frac{\partial}{\partial x_i}e_{kk} \\ &= -\frac{\partial}{\partial x_i}(p - \tfrac{1}{3}\mu\Delta) + \mu\nabla^2 u_i \qquad (15),\end{aligned}$$

where we now write the scalar

$$e_{mm} = \Delta \qquad (16).$$

In time dt the outflow from a volume element $dx_1\,dx_2\,dx_3$ is $\{\partial(\rho u_i)/\partial x_i\}\,d\tau\,dt$. The mass within the element therefore decreases at a rate $\{\partial(\rho u_i)/\partial x_i\}\,d\tau$, and we have the equation of continuity

$$\frac{\partial\rho}{\partial t} = -\frac{\partial}{\partial x_i}(\rho u_i) \qquad (17),$$

or

$$\frac{d\rho}{dt} = -\rho\frac{\partial u_i}{\partial x_i} = -\rho\Delta \qquad (18).$$

Here $\dfrac{d}{dt}$ denotes differentiation with regard to the time following a given particle of the fluid, so that $\dfrac{dx_i}{dt}$ are given by

$$\frac{dx_i}{dt} = u_i \qquad (19),$$

and

$$\frac{d}{dt} = \frac{\partial}{\partial t} + \frac{dx_i}{dt}\frac{\partial}{\partial x_i} = \frac{\partial}{\partial t} + u_i\frac{\partial}{\partial x_i} \qquad (20).$$

In most works on hydrodynamics this operator is denoted by D/Dt, but I see no reason for departing from the usual notation for total differentials, since this particular one is the only total differential that occurs in hydrodynamics. The acceleration components are given similarly by

$$f_i = \frac{du_i}{dt} = \frac{\partial u_i}{\partial t} + u_k \frac{\partial u_i}{\partial x_k} \qquad (21).$$

Consider now the circulation Ω about a closed contour. We have

$$\frac{d\Omega}{dt} = \frac{d}{dt}\int_C u_i dx_i = \int_C \frac{du_i}{dt} dx_i + \int_C u_i du_i$$

$$= \int_C \left(X_i + \frac{1}{\rho}\frac{\partial p_{ki}}{\partial x_k}\right) dx_i + \int_C u_i du_i \qquad (22).$$

The integral $\int_C u_i du_i$ always vanishes; $\int_C X_i dx_i$ vanishes for bodily forces derived from a potential. Then

$$\frac{d\Omega}{dt} = \int_C \frac{1}{\rho}\frac{\partial}{\partial x_k}\{-(p + \tfrac{2}{3}\mu\Delta)\,\delta_{ik} + 2\mu e_{ik}\}\,dx_i$$

$$= -\int_C \frac{1}{\rho}\frac{\partial}{\partial x_i}(p + \tfrac{2}{3}\mu\Delta)\,dx_i + \int_C \frac{1}{\rho}\frac{\partial}{\partial x_k}(2\mu e_{ik})\,dx_i$$

$$= -\int_C \frac{1}{\rho}\,d\,(p + \tfrac{2}{3}\mu\Delta) + \int_C \frac{1}{\rho}\frac{\partial}{\partial x_k}\left\{\mu\left(-2\xi_{ik} + 2\frac{\partial u_k}{\partial x_i}\right)\right\}dx_i$$

$$= -\int_C \frac{1}{\rho}\,d\,(p + \tfrac{2}{3}\mu\Delta) - \int_C \frac{2}{\rho}\frac{\partial}{\partial x_k}(\mu\xi_{ik})\,dx_i$$

$$+ \int_C \frac{2}{\rho}\frac{\partial \mu}{\partial x_k}du_k + \int_C \frac{2}{\rho}\mu\, d\Delta \qquad (23).$$

If $\mu = 0$ this reduces to the circulation theorem of classical hydrodynamics. In many problems of real fluids μ is small and constant, and Δ small. Then the integrals $\int_C \frac{1}{\rho}\,d\,(\frac{2}{3}\mu\Delta)$, $\int_C \frac{2}{\rho}\frac{\partial\mu}{\partial x_k}du_k$, and $\int_C \frac{2}{\rho}\mu d\Delta$ are zero or products of two small quantities, and can be ignored. $\int_C \frac{1}{\rho}dp$ is zero if ρ is a function of p, which is true in many cases, though exceptions arise when the temperature or the composition varies from place to place. In the commonest case, to considerable accuracy,

$$\frac{d\Omega}{dt} = -\int_C \frac{\partial}{\partial x_k} (2\mu \xi_{ik})\, dx_i \qquad (24),$$

and circulation arises only from variations of vorticity in the neighbourhood of the contour. It follows that in a fluid originally at rest or in irrotational motion circulation can arise only through the diffusion of vorticity inwards from a boundary.

The work done on the element $d\tau$ per unit time exceeds the rate of increase of kinetic energy by

$$p_{ki} \frac{\partial u_i}{\partial x_k}\, d\tau = p_{ik} e_{ik} d\tau \qquad (25),$$

and

$$\begin{aligned} p_{ik} e_{ik} &= -\left(p + \tfrac{2}{3}\mu\Delta\right) \delta_{ik} e_{ik} + 2\mu e_{ik} e_{ik} \\ &= -\left(p + \tfrac{2}{3}\mu\Delta\right) \Delta + 2\mu e_{ik} e_{ik} \qquad (26). \end{aligned}$$

But

$$-p\Delta = \frac{p}{\rho} \frac{d\rho}{dt} \qquad (27),$$

and if ρ is a function of p (including a constant as a particular case) this is a differential with regard to the time and yields on the whole zero if the original density is ever recovered. This term may therefore be considered to give the increase of internal energy due to compression. The remainder may be written

$$\begin{aligned} \Phi &= -\tfrac{2}{3}\mu\Delta^2 + 2\mu e_{ik} e_{ik} \\ &= \tfrac{2}{3}\mu \left\{(e_{22} - e_{33})^2 + (e_{33} - e_{11})^2 + (e_{11} - e_{22})^2\right\} \\ &\quad + 4\mu \left(e_{23}{}^2 + e_{31}{}^2 + e_{12}{}^2\right) \qquad (28). \end{aligned}$$

Thus Φ is analogous in form to the distortional strain energy of elasticity ((56) of Chapter VIII). It is essentially positive, and therefore represents work done on the fluid and continually stored as internal energy. In (8), therefore, if the change is an actual one, the term (27) contributes the $p\,\delta\rho/\rho$, while (28) contributes to δQ. So long as the initial and final states are given, it is immaterial whether δQ represents heat conducted into the element, absorbed from radiation, or generated chemically within it, or mechanical energy dissipated into heat by viscosity.

We notice that Φ can vanish if, and only if,

$$e_{11} = e_{22} = e_{33},$$
$$e_{23} = e_{31} = e_{12} = 0 \qquad (29),$$

so that the deformation represents a symmetrical expression or contraction.

If we consider any finite volume, the rate of dissipation within it is

$$\iiint \Phi\, d\tau = \iiint (-\tfrac{2}{3}\mu\Delta^2 + 2\mu\, e_{ik}e_{ik})\, d\tau \qquad (30).$$

But
$$e_{ik}e_{ik} = \xi_{ik}\xi_{ik} + \frac{\partial u_k}{\partial x_i}\frac{\partial u_i}{\partial x_k} \qquad (31),$$

and

$$\iiint \frac{\partial u_k}{\partial x_i}\frac{\partial u_i}{\partial x_k}\, d\tau = \iint l_k u_i \frac{\partial u_k}{\partial x_i}\, dS - \iiint u_i \frac{\partial}{\partial x_k}\left(\frac{\partial u_k}{\partial x_i}\right) d\tau \qquad (32),$$

where l_k is a direction cosine of the outward normal to the boundary. But

$$\iint l_k u_i \frac{\partial u_k}{\partial x_i}\, dS = \iint l_k u_i \left(2\xi_{ik} + \frac{\partial u_i}{\partial x_k}\right) dS$$
$$= \iint 2 l_k u_i \xi_{ik}\, dS + \tfrac{1}{2}\iint \frac{\partial u_i^2}{\partial n}\, dS$$
$$= \iint 2\epsilon_{ikm} u_i l_k \xi_m\, dS + \tfrac{1}{2}\iint \frac{\partial q^2}{\partial n}\, dS \qquad (33),$$

where $\partial/\partial n$ denotes differentiation along the outward normal, q is the resultant velocity, and ξ_m is the vorticity vector associated with ξ_{ik}. Also

$$\iiint u_i \frac{\partial}{\partial x_k}\left(\frac{\partial u_k}{\partial x_i}\right) d\tau = \iiint u_i \frac{\partial \Delta}{\partial x_i}\, d\tau = \iiint \left(\frac{d\Delta}{dt} - \frac{\partial \Delta}{\partial t}\right) d\tau \qquad (34),$$

and in all

$$\iiint \Phi\, d\tau = \mu \iint \frac{\partial q^2}{\partial n}\, dS + 4\mu \iint \epsilon_{ikm} u_i l_k \xi_m dS$$

$$+ 2\mu \iiint \left(-\tfrac{1}{3}\Delta^2 + \xi_{ik}\xi_{ik} + \frac{d\Delta}{dt} - \frac{\partial \Delta}{\partial t}\right) d\tau \quad (35).$$

This form is useful in such a problem as that of waves on deep water, where the viscosity is small; if it were absent we should have a permanent oscillation in a normal mode. The vorticity is negligible except within a distance from the bottom of order $(\nu/\gamma)^{\frac{1}{2}}$, where $\mu = \nu\rho$ and $2\pi/\gamma$ is the period of the motion. Then in the second integral ξ_m is zero at the free surface and u_i is zero at the bottom, so that this integral vanishes. Δ is negligible everywhere. Near the bottom ξ_{ik} contains a factor proportional to the velocity that would exist there on the classical theory, and thus can be made indefinitely small for deep water. Hence the important term in (35) is

$$\mu \iint \frac{\partial q^2}{\partial n}\, dS;$$

and even this vanishes at the bottom, so that it need only be estimated at the free surface.

If a portion of the fluid is compressed without change of temperature, we define the bulk-modulus k by

$$\frac{1}{\rho}\frac{\partial \rho}{\partial p} = \frac{1}{k} \qquad (36),$$

and if it expands under change of temperature without

change of pressure we define a coefficient of volume expansion α by

$$\frac{1}{\rho}\frac{\partial \rho}{\partial \theta} = -\alpha \qquad (37).$$

Then for small changes of temperature and pressure

$$\delta\rho = \rho\left(\frac{\delta p}{k} - \alpha\,\delta\theta\right) \qquad (38).$$

Also in a free expansion under constant pressure

$$\begin{aligned}\delta Q &= M\delta\rho + \rho c\,\delta\theta \\ &= \left\{\theta\,\frac{\partial}{\partial\theta}\left(\frac{p}{\rho}\right).\,\alpha\rho + \rho c\right\}\delta\theta \qquad (39),\end{aligned}$$

where the partial differentiation is to be carried out at constant density. But this makes

$$\frac{\partial p}{\partial \theta} = \alpha k \qquad (40),$$

and therefore $$\delta Q = \rho\left\{c + \alpha^2\frac{k}{\rho}\,\theta\right\}\delta\theta \qquad (41),$$

so that c_p, the specific heat at constant pressure, is given by

$$c_p = c + \frac{\alpha^2 k\theta}{\rho} \qquad (42).$$

This form is analogous to that of (56) of Chapter VIII, the present α being the coefficient of volume expansion, and the previous one that of linear expansion.

In an adiabatic change $\delta Q = 0$; then

$$-\,\alpha k\theta\,\frac{\delta\rho}{\rho} + \rho c\,\delta\theta = 0 \qquad (43),$$

and therefore $$\begin{aligned}\alpha\theta\,\delta p &= (\rho c + \alpha^2 k\theta)\,\delta\theta \\ &= \left(1 + \frac{\alpha^2 k\theta}{\rho c}\right)\frac{\alpha k\theta}{\rho}\,\delta\rho,\end{aligned}$$

so that $$\delta p = k\left(1 + \frac{\alpha^2 k\theta}{\rho c}\right)\frac{\delta\rho}{\rho} \qquad (44).$$

Thus the bulk-modulus for adiabatic changes is

$$k' = k\left(1 + \frac{\alpha^2 k\theta}{\rho c}\right) = \frac{kc_p}{c} \qquad (45).$$

The equation of heat conduction needs to be modified to allow for heat generated internally. In time δt, per volume $d\tau$, we have

$$\delta Q = \Phi\,\delta t + \frac{\partial}{\partial x_i}\left(K\,\frac{\partial\theta}{\partial x_i}\right)\delta t$$

$$= \rho c\,\delta\theta + \alpha k\theta\,\Delta\,\delta t \qquad (46),$$

whence

$$\rho c\,\frac{\partial\theta}{\partial t} + \alpha k\theta\Delta = \Phi + \frac{\partial}{\partial x_i}\left(K\,\frac{\partial\theta}{\partial x_i}\right) \qquad (47).$$

EXAMPLES

1. Obtain the equations of motion in terms of the stress components by considering the momentum of a finite volume of any form and applying Green's theorem.

2. Similarly obtain equation (25).

INDEX